Havant College LRC
Withdrawn
D0298190

Tourism and Political Boundaries

More people are travelling now than ever before, and few places exist in the world today that have not been penetrated by tourism. Every year millions of people travel across political boundaries seeking encounters with cultural and natural environments different from their own. Borders not only define the differences between the origins and destinations of tourists, but also fashion travellers' impressions of place and leisure experiences.

Based on a wealth of empirical examples from around the world and concepts borrowed from a wide range of disciplines, *Tourism and Political Boundaries* provides an accessible, comprehensive glimpse into the relationship between tourism and borders.

The subjects covered include:

- The formation of political boundaries in the context of tourism, and the meanings behind these boundaries.
- The barrier effects of borders in relation to the flow of tourists and the development of tourism.
- Political boundaries as tourist attractions/destinations.
- The ways in which tourism landscapes are created in the presence of political frontiers.
- Global transformations occurring in the realm of geopolitics and their influence on tourism.
- Special tourism planning considerations in border regions.

This book provides innovative information for scholars who are interested in globalisation processes, particularly within the context of tourism. The political, economic, spatial and psychological implications of the boundary-tourism nexus will be of interest to students and researchers from a variety of intellectual backgrounds.

Dallen J. Timothy is Assistant Professor at Arizona State University, USA. He has published extensively in the areas of tourism planning, developing regions, heritage and political boundaries and tourism. He is author and co-editor of *Women as Producers and Consumers of Tourism in Developing Regions* (Praeger), *Tourism in Destination Communities* (CABI) and *Heritage Tourism* (Longman).

Routledge Advances in Tourism
Series Editors: Brian Goodall and Gregory Ashworth

1 The Sociology of Tourism
Theoretical and empirical investigations
Edited by Yiorgos Apostolopoulos, Stella Leivadi and Andrew Yiannakis

2 Creating Island Resorts
Brian King

3 Destinations
Cultural landscapes of tourism
Greg Ringer

4 Mediterranean Tourism
Facets of socioeconomic development and cultural change
Edited by Yiorgos Apostolopoulos, Lila Leontidou, Philippos Loukissas

5 Outdoor Recreation Management
John Pigram and John Jenkins

6 Tourism Development
Contemporary issues
Edited by Douglas G. Pearce and Richard W. Butler

7 Tourism and Sustainable Community Development
Edited by Greg Richards and Derek Hall

8 Tourism and Political Boundaries
Dallen J. Timothy

9 Leisure and Tourism Landscapes
Social and cultural geographies
Cara Aitchison, Nicola E. MacLeod and Stephen J. Shaw

10 Tourism in the Age of Globalisation
Edited by Salah Wahab and Chris Cooper

Tourism and Political Boundaries

Dallen J. Timothy

London and New York

First published 2001
by Routledge
11 New Fetter Lane, London EC4P 4EE

Simultaneously published in the USA and Canada
by Routledge
29 West 35th Street, New York, NY 10001

Routledge is an imprint of the Taylor & Francis Group

Typeset in Baskerville by Keyword Publishing Services Ltd
Printed and bound in Great Britain by University Press, Cambridge

British Library Cataloguing in Publication Data
A catalogue record for this book is available from the British Library

Library of Congress Cataloging in Publication Data
Timothy, Dallen J.
Tourism and political boundaries / Dallen J. Timothy.
p. cm.
Includes bibliographical references (p.).
1. Tourism. 2. Boundaries. I. Title.

G155.Al T535 2001
338.4′791–dc21 00-067289

ISBN 0–415–19696–5

To Carol, Kendall, Olivia, Aaron, and Spencer
and to the memory of Arden Jay Timothy (1938–97)

Contents

Plates

Figures

Tables

Preface

The origins of this book are from my childhood. My grandmother, Fern DeMille, with her tales of exotic lands and her deep-seated academic curiosity of foreign places stemming from her own heritage and her inquisitive nature fostered a contagious and unquenchable thirst for learning about, and experiencing, distant places. While opportunities to travel overseas came to her only in her later years, she regularly gathered her children and grandchildren as youth into her living room and presented family nights filled with lessons of cultures, languages, foods, dress, posters, and material cultural objects – anything she could get her hands on from the limited resource pool of a small desert community. While Grandma Fern is to blame for my abiding desire to traverse the borders of the world, my parents, Arden and Chyrrel Timothy, were, not without trepidation, supportive of my leaving home as a youngster on several occasions to fulfill my dreams.

So, decades later, here I am writing a book on tourism and borders. My travels now have led me to 81 countries, and I never cease to be amazed at the political, social, and economic functions of international boundaries. For many, they are so simple and insignificant, but for me, and hopefully others, they represent a bountiful area of research that helps me in my quest to learn of places foreign. Despite the fact that hundreds of millions of people cross national frontiers every year, the links between borders and tourism are not well defined, and the subject has been all but completely ignored by scholars of both tourism and borders. From this awareness, the idea for this book was born. The theories, concepts, and empirical examples presented here are based on four primary sources: (1) the popular media, (2) academic literature, (3) my formal research projects along the borders of North America, Africa, Europe, and Asia, and (4) my own experiences from the past 30 years of crossing, carefully observing, photographing, straddling, and thinking about borders.

My goal is for scholars to realize the importance of boundaries/frontiers/borders in international travel and begin to devote increased attention to understanding them. I also sincerely hope that travelers will stop, think and examine the meanings and functions of the boundaries they cross.

There is much to learn about the human experience, and borders can play a major role in that, particularly since they demonstrate clearly the extremes of socio-political life in microcosmic form.

Dallen J. Timothy
Gilbert, Arizona
August 2000

Acknowledgments

Several people have been influential in the development of ideas and concepts for this book. I was fortunate to have opportunities in my formative years to work closely with Lloyd Hudman, Charles Whebell, Dick Butler, and Geoff Wall, who with all their inexhaustible expertise, patience, and constructive guidance, have inspired in me a deep interest in the geography of tourism. I have also spent countless hours absorbing knowledge from Stephen Boyd, Barbara Carmichael, Ross Dowling, David Groves, Michael Hall, Atsuko Hashimoto, Dimitri Ioannides, Julie Lengfelder, Alan Lew, Bruce Prideaux, Dwita Rhami, Darren Scott, Bakti Setiawan, and David Telfer. All have been excellent colleagues, and thanks are due for allowing me to hold them captive when I felt like talking borders.

To my family, Carol, Kendall, Olivia, Aaron, and Spencer, I owe a special debt of gratitude for allowing me so many late nights and so much understanding when I was in over my head. They give me cause to enjoy coming home every day. Likewise, my father, Arden Timothy, who died too young, deserves special attention for allowing me, even as a youngster, to drag him along on my borderland explorations.

Brian Goodall and Greg Ashworth, the series editors, expressed enthusiasm for this book from the beginning. Thank you for your support. The editorial staff members at Routledge have also been outstanding. My sincere thanks go out to Casey Mein, Craig Fowlie, Joe Whiting, and Simon Whitmore for their patience and for their help in making this work a reality.

Finally, the author and publisher would like to thank the following for permission to reproduce figures in the text: Chad Emmett for permission to use his photograph in Plate 2.1; Nathan Richardson for permission to use his photograph in Plate 3.4; Elsevier Science for permission to reproduce Figure 3.3; Elsevier Science for permission to reproduce Figure 4.2; University of Arizona Press for permission to reproduce Figure 4.3.

1 Borders and tourism

> I love border crossings. They somehow make me feel as though I'm in a black-and-white movie with subtitles
>
> (Harris 1997: 12)

Introduction

Tourism is quickly becoming the world's largest industry and more people are traveling now than ever before. Most basic definitions of tourist include some element of traveling away from one's home environment. When operationalized in this way, tourism almost always involves the crossing of some political boundary. Sub-national boundaries, such as those between provinces and counties, may have significant implications for tourism, especially in terms of planning, promotion, and taxation. International boundaries, however, influence tourism in many more ways. The flow of tourists, their choice of destinations, planning and the physical development of tourism, and the types and extent of marketing campaigns are all affected by the nature of political boundaries.

For centuries travelers have crossed international boundaries for trade, as well as personal enjoyment and education (e.g. the Grand Tour). Despite the significance of borders, and humankind's long history of foreign travel, very little has ever been written, and thus little is known, about them in the context of tourism. Only recently have scholars started to merge border research with tourism, which likely reflects the relative infancy of tourism as an area of academic study and the paucity of political geographers who incorporate their interests into tourism research.

One of the earliest endeavors to examine the relationships between borders and tourism occurred when the International Geographic Union's recreation and tourism study group sponsored conference sessions in 1977 on tourism and borders, and proceedings were subsequently published (Gruber *et al.* 1979). Since that time, only a few writers have begun to consider the political, social, economic, cultural, and psychological effects of borders on tourism. Others have inadvertently stumbled onto the subject and buried it deeply within broader studies of economic development and politics, far from the mainstream of tourism research literature and scattered throughout many discipline- and country-specific media.

This book aims to bring together the scattered concepts and theories that help explain the relationships between borders and tourism into a single volume that will contribute to a better understanding of the subject in the mainstream tourism literature. In particular, this chapter discusses the nature of political boundaries and the development of international tourism. It also lays the groundwork for the rest of the book by examining the relationships between tourism and borders.

Understanding borders

Some scholars have traced the concept of territorial boundaries back to the Roman Empire in reference to the walls constructed in Northern England, Scotland, and along the Rhine and Danube rivers. Critics, however, argue that these were not fixed lines of jurisdiction, because the limits of the Roman State were constantly changing through conquests and because Rome did not recognize the existences of surrounding polities since it saw itself as the 'universal state' (Herzog 1990: 17). Hence, they were not formal lines of administrative jurisdiction, as the concept of borders is understood today. Rather, they functioned as edges of military activity against warring Germanic and Celtic tribes, as well as landscapes that defined an approximation of the empire's sphere of influence (Herzog 1990). The same is true of the Great Wall of China in its role as line of military defense against nomadic barbarians and approximate limit of central influence and control (Prescott 1987).

Throughout the Middle Ages, governments and monarchies controlled territories that were fluid and ill-defined. Most areas of control were separated by undifferentiated zones of physical landscapes (i.e. frontiers), such as swamps, forests, mountain ranges, and vast deserts (Kristoff 1959). National territory was often vaguely defined by the extent of land ownership by aristocracy or as the hinterland of primary cities, and rarely were the frontiers considered indubitable lines of authority. Instead, they were regions of nebulous power that often separated warring empires and into which nationalist expansion could occur.

However, during the nineteenth century, rapid expansion from core areas into the frontier led to the development of nation-states, as the concept of sovereignty became more closely linked to territorial control (Gottman 1973; Johnston 1982; Sack 1986). Thus, the idea of fixed boundary lines between sovereign authorities began to develop as states became increasingly viewed as the ultimate source of legitimate authority and law (Herzog 1990). This parallels LeFebvre's (1991) notion of absolute space – easily recognized territory that is defined by boundaries and legal descriptions. States thus have authority to manage the phenomena, processes, and activities occurring within their bounded space.

For many years political geographers and other social scientists have directed their attention to various aspects of political space and action. Political boundaries have been one area of primary concern among scholars for more than a century (e.g. Ratzel 1892; 1896), with particular emphasis being placed upon descriptions of how borders throughout the world were created and frontier regions as zones of conflict (Minghi 1963a). These strategic themes gained momentum during and between the first and second world wars (Boggs 1940; Holdich 1916; Jones 1945).

Around the middle of the twentieth century scholars began to realize that there was more to the study of borders than just describing the types of artifacts used to demarcate them and chronicling the locations and causes of territorial disputes. It became clear that borders also exerted significant influences on the economic and sociological aspects of the human experience (do Amaral 1994; Minghi 1963b), which created distinctly different spatial and temporal patterns on opposite sides. Thus began an evolution in border studies from a linear focus concerned primarily with the physical structures of the border to an areal focus concerned more with broader economic and socio-cultural implications of the border.

More recently, scholars have begun examining 'borders of the mind' – the psychological effects of borders and the deeper underlying meanings that societies and governments ascribe to them (Herzog 1990; House 1980; Minghi 1994a; Rumley and Minghi 1991). This is manifested in an increased interest in the study of border regions and borderlands, as researchers now recognize that boundaries have depth, or spheres of influence, and that they have different meanings for different groups of people. In fact, as Knight (1994) reiterates, as with all regions that pertain to human social organization, boundaries and state territory are a social construct. Similarly, according to Knight (1982: 517) territory and boundaries are not; they become, 'for territory itself is passive, and it is human beliefs and actions that give territory meaning.'

Border regions, which Hansen (1981: 19) defined as 'areas whose economic and social life is directly and significantly affected by proximity to an international boundary', are the focus of this book, although reference will be made to the broader political realms within the state. Although some authors have identified differences between border regions and borderlands (e.g. McKinsey and Konrad 1989), these terms will both be used to refer to the areas adjacent to boundaries on both sides. Likewise, despite the differences in meaning of frontiers and borders, as discussed earlier, the terms are used interchangeably throughout this book.

Scales of borders

Various scales, or levels, of boundaries exist, and each of them has its own purpose. National, or international, boundaries comprise the first level of political control. National boundaries have the most obvious impacts on the natural environment, economic operations, and patterns of socio-cultural interaction. At international frontiers currencies change from francs to marks and bahts to kyats, and in the case of the United States and its neighbors, distances change from miles to kilometers. Some borders even define an abrupt change in language, religion, political attitudes, cultural traditions, and social mores. National holidays on either side of a border are obviously different and business hours may vary.

Sub-national boundaries, such as those between states, provinces, cantons, and departments, can be viewed as second-level boundaries. These too can have significant effects on the human experience. Education levels and law enforcement policies may differ, and quality of infrastructure maintenance can vary. In the United States and Canada, sales taxes, driving ages, drinking ages, and gambling laws often change from state to state and province to province.

Third-order civil divisions include counties, townships, and municipalities. These lower-level frontiers have the fewest impacts on human interactions, but they are nonetheless significant. Such borders may influence and determine property tax rates, law enforcement procedures, and insurance coverage.

In addition to 'normal' political borders, other types of special-purpose administrative areas exist, which have their own boundaries, but which do not fit within the hierarchy described above. Indian reservations in the United States are examples of a unique type of territory that has been granted autonomy over several areas of governance, such as taxation, legislative control, law enforcement, education, and economic development (Eadington 1990; Lew and Van Otten 1998; Pommersheim 1989). This high level of self governance, which is mandated at the national level and outside the control of the states in which they are located, has allowed the growth of several forms of tourism (e.g. gaming, cultural, and nature-based) on native reserves in North America.

National, provincial, and state parks are also examples of other special-purpose territories whose borders have been established by various levels of government for the conservation of natural and cultural environments. Sometimes park boundaries are the focus of major disputes between public agencies and local residents (Morehouse 1996), and they determine the spatial patterns and types of development that can occur in some areas of a country.

Types of boundaries

Consensus among early scholars was that the best political boundaries are those that follow natural features, such as rivers and crests of mountain ranges (Hartshorne 1936; Holdich 1916; Jones 1943; 1945) because they are more permanent than human-created lines and population regions tend also to separate along those features. This, they suggested, will decrease problems associated with bisecting like cultural groups with artificial political divides. According to Glassner (1996: 89), however, no type of boundary is better than any other, except of course the type that performs the fewest functions – the one 'falling between good neighbors.'

All boundaries are human creations, even physiographic ones since nature rarely, if ever, draws clear border lines (Leimgruber 1991), and humans inevitably impose cultural and political values on nature.

> It is not natural to divide drainage basins, dissect common plains, split mountain ranges, or divide up surface and underground water resources. Yet this is precisely what boundary lines do ... boundaries do not only cut across natural resources, they also impede human mobility and transaction, and many do not respect ancient tribal boundaries, linguistic borders, ethnic groupings or the cultural landscape.
>
> (Grundy-Warr and Schofield 1990: 11)

Geometric boundaries are those which do not necessarily correspond with natural features, but instead consist of human-created line segments that are measured without physiographic rationale. These are usually based on lines of latitude and longitude or cultural features, such as roads or edges of human settlements.

In 1936 Hartshorne suggested what Glassner (1996: 89) labeled a genetic classification of political boundaries based on when they were established in relation to human settlement. According to Hartshorne, an antecedent boundary is one that precedes the development of most features of the cultural landscape. A boundary that is completely antecedent, or found where the line was drawn prior to human settlement, is known as a pioneer boundary (e.g. United States–Canada west of the Great Lakes). Many boundaries are subsequent to human settlement and therefore superimposed on the existing cultural landscape, regardless of what lies in the way (e.g. most African and European boundaries). Relict boundaries are those that no longer function as borders but are still visible in the cultural landscape (e.g. parts of the Berlin Wall and the Great Wall of China).

Process of development

Jones (1945) suggested that ideally, and under 'normal' conditions, the establishment of international boundaries should undergo a series of four steps. First, boundaries are defined; that is, they are described in relation to the terrain through which they run. This description uses natural features, such as rivers and hills, as well as cultural features, such as roads, fences, and buildings, to identify the location of the boundary as precisely as possible. The second step, delimitation, refers to the process of plotting on air photos and topographic maps the exact location of the border. Third, the border line is then marked on the ground with the aid of survey equipment, based on the original definition and delimitation documents. This process is known as demarcation. Methods of demarcation range from simple poles and rock piles to complex systems of fences and walls adorned with barbed wire. The way a border is demarcated reveals a great deal about the nature of the relationship between adjacent countries. The final step is maintenance, or administration, which refers to the process of upkeep and maintenance of the boundary markers and the enactment of legislation that deals directly with the boundary itself.

While this is an ideal pattern of development, few boundaries throughout the world were created in this manner. The meaning of 'normal' conditions is debatable, since most boundaries have been founded by unilateral aggression or are merely remnant property lines from medieval days when gentry ownership defined national territory. Clearly, even with the best surveying equipment, delimitation and demarcation are not an exact science. Many border skirmishes have resulted from miscalculated boundary markings, and most of the ongoing battles of today are based on territorial and border problems.

Boundary functions

Several authors have described the functions of political boundaries (Leimgruber 1980; Pearcy 1965; Prescott 1987). Five primary functions are most notable. First, borders are legal limits that define the territory of a state; they are the lines up to which a political entity can exercise its sovereign and legal authority. No matter how close people live to the border, they are subject to the laws in place on their side of the boundary (Glassner 1996), even though political unrest might make legal control difficult. Furthermore, governments have the right to collect taxes from residents dwelling within national territory, establish education standards, and implement development programs. Second, boundaries play an important economic role. By filtering the flow of goods across national frontiers or levying high tariffs and duties, nations can protect their economies by limiting domestic competition with foreign producers. Filter mechanisms are also created to keep undesirable elements out, such as drugs, weapons, endangered species, and foodstuffs that are feared to be contaminated by disease, or desired elements in, such as large sums of currency and gold. It is generally the duty of customs agencies to control the flow of goods. The third function is to monitor and control the flow of people. Some countries erect strong physical and legal barriers to keep certain undesirable people out or to limit the number of people that can enter. Illegal immigrants, criminals, and people with certain health problems are commonly targeted. The United States, for example, has erected a steel and cement barrier along its land border with Mexico in an effort to thwart illegal migration from Latin America. Other countries erect similar fortifications to keep people from leaving. The East–West Germany border prior to 1989, for example, was marked with fences, walls, minefields and guard towers, all in the name of keeping Eastern Germans from escaping to the West. Fourth, ideological barriers are created when a country enacts strict regulations to prevent the drift of ideas and information across boundaries. During the Cold War era, for instance, religious literature was not permitted to enter Eastern Europe, and communist governments made every effort to obstruct communications from the West. Finally, some borders function as lines of military defense. In some areas (e.g. Israel–Lebanon and North–South Korea), borders are heavily fortified in an effort to keep armies and terrorists out and to exert extraterritorial claims to land in adjacent countries.

According to Prescott (1987: 80), the only real boundary function is to mark the limits of sovereignty; the other elements discussed above are merely residual functions that are placed at national frontiers as needs arise. Minghi (1963a) supports this claim by suggesting that border functions are not static but do change over time, and Boggs (1932) suggested early on that boundary problems which cause friction between neighbors might better be solved by revising the functions assigned to a border than by modifying the border itself – advice that is rarely heeded in a world where every meter of national territory is jealously guarded.

Human interaction across international boundaries

As long as boundaries have existed, human interaction across them has been of concern to national governments. As mentioned above, one function of boundaries is to hinder and monitor the flow of people and other forms of interaction into and out of a country. The degree of permeability depends on the functions of the boundary in question as well as the degree of sociocultural similarity on each side (Ante 1982; Donnan and Wilson 1999). Despite rapid political changes globally, unfavorable international relations in some areas still keep neighbors in a state of limited interaction (House 1980; 1981; Martinez 1994). Several authors have suggested that it is also important to consider cultural and linguistic similarities and differences when considering human interaction. Many people will be more inclined to cross the border if the language on the other side is the same as their own (Eriksson 1979; Leimgruber 1981; Timothy 1995b). However, adventurous travelers would more likely be attracted to different cultures on opposite sides and consider this a primary reason for crossing. Regardless of the existence of obstacles to cross-border interaction, international trips number in the hundreds of millions each year.

Tourism

In accordance with World Tourism Organization (WTO) definitions, international tourists are people who cross an international boundary and stay at least one night in the destination country. Most of the statistics collected by the WTO reflect this definition. As illustrated in Table 1.1, these figures are substantial, but do not even take into account the millions more people who cross national borders and return home the same day (World Tourism Organization 2000b). International day trips in both directions along the US–Mexico and US–Canada borders alone account for over one hundred million additional and uncounted international trips every year.

Nonetheless, as Table 1.1 shows, for over 50 years, tourism has grown at a steady rate. In 1950, approximately 25 million international overnight trips were taken. This number grew to 457 million in 1990, but in 1999 had increased to 663 million, an increase in only ten years of an astounding 45 percent. Growth slowed in 1991 largely as a result of the Gulf War and again in 1997–98 consequent to the Asian financial crisis. But, in 1999 international overnight arrivals increased by 6 percent from the year before, generating approximately US$4 trillion of economic activity. According to WTO forecasts, tourism is expected to expand by an average of 4.1 percent a year over the next two decades, surpassing a total of one billion international trips by the year 2010 and reaching 1.6 billion by the year 2020 (WTO 2000b).

Global tourism is spatially concentrated (Pearce 1999a). The ten leading destinations, for example, account for 52 percent of all international arrivals. In 1998, the top 20 countries accounted for 71 percent of all world arrivals. Table

Table 1.1 International tourist arrivals

Year	*Arrivals (millions)*	*Percent of growth*
1950	25	—
1960	69	176
1970	166	141
1980	288	73
1985	330	15
1990	457	38
1991	463	1
1992	503	9
1993	520	3
1994	551	6
1995	565	3
1996	596	6
1997	613	3
1998	625	2
1999	663	6

Source: World Tourism Organization 2000a

1.2 lists the world's leading 20 destination countries. By itself France possesses over 11 percent market share of the world total, followed at a fair distance by Spain and the United States with 7.6 percent and 7.5 percent respectively.

International travel has existed for millennia as explorers, traders, and religious pilgrims traveled in search of conquest, commodities, and spiritual enlightenment. Travel for pleasure and education, however, began in the 1700s among the wealthy in Europe following the Seven Years' War and the beginning of the French Revolution, a time when moving through France and Italy could be done in relative safety (Lundberg and Lundberg 1993). With every successive generation, however, tourism has moved from being exclusively for the rich to being accessible to the masses. Patterns of tourism growth parallel the increase in levels of global affluence. In developing countries, for example, domestic tourism is rising as improved standards of living have meant that domestic travel is becoming more accessible to a wider cross-section of society (Pearce 1999a: 1). International travel is also growing in importance among members of developing societies as more people have access to opportunities for traveling to neighboring countries. This is particularly the case in Southeast Asia, where intraregional travel has grown dramatically in recent years (Hall and Page 2000). Increased levels of afflu-

Table 1.2 The top 20 tourist destination countries by international arrivals, 1998

Rank	*Country*	*International arrivals (thousands)*	*Market share (% of world total)*
1	France	70,000	11.2
2	Spain	47,743	7.6
3	United States	47,127	7.5
4	Italy	34,829	5.6
5	United Kingdom	25,475	4.1
6	China	24,000	3.8
7	Mexico	19,300	3.1
8	Poland	18,820	3.0
9	Canada	18,659	3.0
10	Austria	17,282	2.8
11	Germany	16,504	2.6
12	Czech Republic	16,325	2.6
13	Russian Federation	15,810	2.5
14	Hungary	14,660	2.3
15	Portugal	11,800	1.9
16	Greece	11,077	1.8
17	Switzerland	11,025	1.8
18	China–Hong Kong SAR	9,600	1.5
19	Turkey	9,200	1.5
20	Thailand	7,720	1.2
World total		625,236	100.0

Source: World Tourism Organization 2000a

ence and the rapid development of transportation and computer technology have created a smaller world where few places remain untouched by tourism, and more people are traveling now, for business and pleasure, than ever before.

Tourism and borders

In common with all other types of economic, socio-cultural, and environmental activities, tourism is affected by the existence of political boundaries. Border-related policies, differences in administrative structures on opposite sides, and the physical barrier created by borders can affect many aspects of tourism, including travel motivations and decision making, infrastructure development, marketing and promotion, and place image.

In light of these impacts, Matznetter (1979: 67) highlighted some of the connections between boundaries and tourism, and suggested a three-fold typology of spatial relationships between the two: (1) where the boundary line is distant from tourist areas, (2) where a tourist zone exists adjacent to the boundary on only one side, and (3) tourist zones that extend across, or meet at, borders. He suggests that in the first case, the frontier functions as a barrier or simple line of transit. Thus, the influence of the border depends largely on its degree of permeability. In the second case, Matznetter suggests that in addition to being attracted to the tourist-oriented side, some people will be attracted to visit the other side as well, which presents opportunities for tourism development to spill onto the non-tourist side of the border. In the third instance, there may be communication and cooperation between the two sides, so that the entire natural or cultural attraction system operates as one entity, or conversely the border may act as a significant barrier altogether. While Matznetter did not consider all possible spatial arrangements of borders and tourism, and although he only considered a few of the impacts that borders have on tourism, his paper was a useful attempt at conceptualizing tourism and international boundaries.

This book builds upon Matznetter's (1979) efforts by examining several of the relationships between tourism and political boundaries, the causes and effects, as well as the symbiotic relationships that exist. While the real implications of the intersection between borders and tourism are multifaceted, they will be examined in this book within the following conceptual framework, which has its foundations in the border research literature: borders as barriers, borders as destinations, and borders as modifiers of the tourism landscape (Figure 1.1). Boundaries form real and perceived barriers to international travel owing to specific functions and methods of demarcation, as well as the experiences and expectations of individual travelers concerning the border itself or what lies on the opposite side. Various scales of borders also act as tourist attractions when they offer a unique spectacle in the cultural landscape. The depth and influence of the frontier into the borderlands

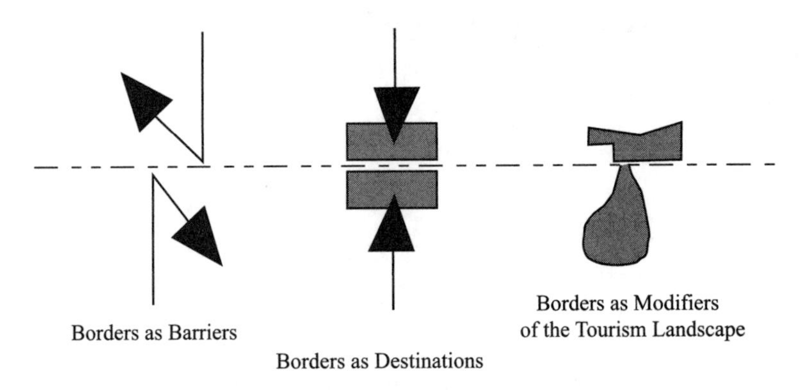

Figure 1.1 The relationships between borders and tourism

also create economic, legal, and cultural differences that become significant attractions in many locations. Finally, human landscapes unique to border regions also strongly influence the way tourism develops spatially on opposite sides of a border – not simply in the borderlands but also deep within national space.

These three primary relationships (chapters 2–4) comprise the more traditional views of political boundaries in that they focus more on the cultural landscapes, conflicts, barriers, and push and pull factors that are commonly attributed to boundary locations. However, this picture is not complete. The book also attempts to answer questions of change (chapter 5) as monumental political transitions have occurred throughout the world during the past 15 years and as the process of rapid globalization continues to race on, resulting in clear alterations of the more traditional functions and roles of political frontiers. Furthermore, as an extension to this, chapter 6 examines the implications of these border-related changes on tourism planning for borderland regions, which, in addition to the difficulties faced by planners everywhere, experience additional challenges owing to their peripheral and boundary locations. The final chapter summarizes the main ideas brought out in the book and proposes additional areas of research that might be undertaken by social scientists who are interested in the relationships between boundaries (political and other) and tourism.

2 Borders and barriers

> For some people [crossing borders] is merely annoying, but for others it can be frightening
>
> (Budd 1990: 15)

Introduction

There is widespread recognition that political boundaries create barriers to human interaction. Borders are common sore spots in international conflicts, and restrictive government policies are often most vivid in frontier communities and at border crossing points. Service industries, such as tourism, are particularly sensitive to political frontiers and their associated formalities and problems (Tucker and Sundberg 1988).

International boundaries can be viewed as barriers to travel from at least two perspectives: real and perceived. Real barriers create insurmountable constraints to tourism because they either hinder tourist flows physically or, through strict border-related policies, make travel difficult or virtually impossible. Perceived barriers do not generally pose real physical obstacles to border crossing. Instead, they create conditions wherein border crossing is challenging and therefore undesirable. Thus, people are permitted to cross, but owing to perceived constraints, do not. In addition to hindering the flow of tourists, the planning, development, and promotion of tourism in destination areas can be significantly hampered in the face of border conflicts and other political problems. This chapter examines the barrier effects of boundaries in relation to tourist flows and the growth and development of tourism.

Real barriers

As mentioned in chapter 1, one of the primary functions of international boundaries is to control the flow of people in and out of a country. This can be done in at least two ways. First, heavy fortification and defensive demarcation methods physically keep people in a country, which was of primary importance along the former East–West German frontier. They can also keep people out, such as in the case of the walls and fences along the US–Mexico border, erected by the United States to keep illegal immigrants from crossing. Second, strict frontier-crossing formalities can be operationalized by home and host country that will function as a filter to keep people at home or to keep undesirables out.

International borders possess different degrees of permeability, ranging from open crossings with no checkpoints to borders that are completely closed and which no one is permitted to cross. An example of the former type would be the Liechtenstein–Switzerland frontier, and an example of the latter would be the border between North and South Korea. Figure 2.1 demonstrates a number of examples along this spectrum of permeability. This condition depends to a large degree on the friendliness between nations and the history of how their common boundary was established. Real barriers, such as those associated with strict formalities and defensive demarcation methods, necessarily deter many people from crossing a border. Indeed this may be one of the primary aims of such a boundary (Timothy 1995a; 1998c).

This notion of permeability varies from place to place, however, sometimes depending on which side of a border a person lives. When physical demarcation and strict border policies work together, however, which they usually do in cases of highly fortified boundaries, the border becomes even more impermeable. For example, the walls and fences along the US–Mexico border are the same for residents of both nations, but they are much more permeable for Americans than for Mexicans, owing to strict legal controls on non-Americans entering the country.

Political problems and borderland crime also act as real barriers to tourism in terms of affecting the flow of tourists and the development of the industry itself. Wars, border disputes, and crime can physically destroy natural and cultural resources and prevent the successful development of tourism in certain destinations. The same problems also deter many potential tourists from visiting disturbed areas.

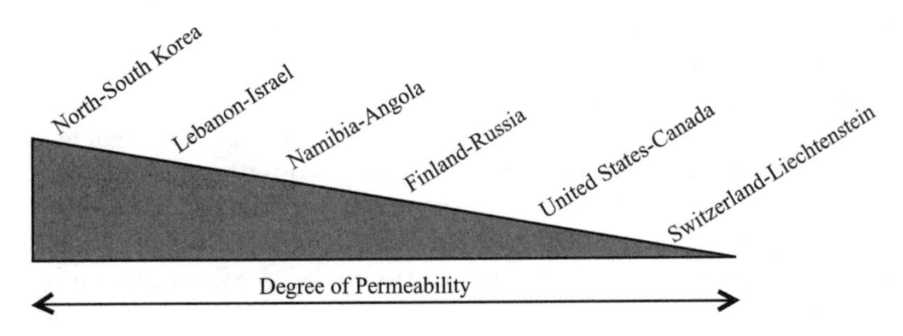

Figure 2.1 Boundary permeability

Unfavorable international relations

In addition to the physical manifestations of negative relations between countries discussed above (e.g. physical barriers), unfavorable international relations also create severe restrictions on travel between adversarial countries that are in fact as much real barriers as are the barbed-wire fences and fortified walls described earlier. Following the Turkish intervention in northern Cyprus in 1974, which will be discussed later in this chapter, relations between Turkey and Greece diminished rapidly. As a result of these unfavorable associations between the two sides, the Greek Cypriot government has established restrictions on travel to and from the north. Travelers who enter the island in the north are not permitted to enter the south. Tourists who enter the island in the south are allowed to visit the Turkish north, but only during the day, and they may not stay overnight, or bring back items purchased in the north (Akis and Warner 1994; Kliot and Mansfeld 1994; Lockhart and Ashton 1990). Furthermore, the government of the south has declared the ports and airports in the north illegal points of entry, and maps of Cyprus produced in the south depict the north as inaccessible to visitors (Akis and Warner 1994).

Another manifestation of this phenomenon is the policy of most Middle Eastern nations, with the exception of Egypt and Jordan, to refuse entry to travelers who have visited Israel, as is evident from entry and exit stamps in their passports. Another example is the US government's laws against allowing US citizens to travel to Cuba, Libya, or Iraq without special permission from the Department of the Treasury. US passports carry a warning to this effect. In 1996, 43,000 Cuban Americans and 17,000 other qualified visitors were given legal permission by the US government to travel to Cuba (Reynolds 1997). The island is increasing in popularity as a tourist destination among Canadians and Europeans and would no doubt be popular among Americans if they were permitted to travel there. In addition to those legally allowed to visit Cuba, thousands of other American tourists skirt US government restrictions every year by traveling to Cuba either through Canada, Mexico, the Bahamas, or Jamaica. Cuban officials typically do not stamp US passports, so that there is no official evidence of a visit. If Americans are caught, they are generally fined and required to agree not to go there again (Reynolds 1997).

Unfavorable relations between neighboring countries also can force tourists to travel to some third party country in order to get to a neighboring country. The sealing of the Spain–Gibraltar border in 1969 led to the need for travel between Spain and the British colony to be undertaken through a third country, usually Great Britain or Morocco. The closing of the border between Tanzania and Kenya in 1977 forced tourists visiting both countries on the same trip to take a detour via Ethiopia, the Seychelles, or Zambia to reach the neighboring country (Gruber 1979).

Partitioned states, sometimes referred to as divided or quasi states, are separate political units that once were part of a larger political entity (Butler and Mao 1995). Examples of partitioned states include Macau and Taiwan in relation to China, North and South Korea, and Turkish northern Cyprus and the Republic of Cyprus. Some previously partitioned states have reunited, such as Vietnam and Germany. Several forces have historically acted to create partitioned states, such as war, religious differences, colonization, and ethnicity (Waterman 1987). The political boundaries and territorial integrity of partitioned states are not clearly defined because partitioned states are often not recognized as countries by the other units or by the international community. Boundaries between these divided states are rarely considered 'normal.' Rather, they are usually ceasefire lines, demilitarized zones, or other lines of military control (Butler and Mao 1995; 1996). In nearly all cases, partition has been a result of some form of negative political relations, and therefore needs to be examined in this context.

Butler and Mao (1996) explain how tourism between partitioned states and between other states and partitioned states differs from normal patterns. In terms of tourist flows, tourism is classified into either domestic or international. According to Butler and Mao, however, this typology is far too limited when partitioned states are being considered. Travel between the parts of partitioned states is usually not systematically recorded because such travel is commonly restricted, and the partitioned states themselves often do not recognize it (Butler and Mao 1996: 27). This results in a unique situation. Travelers between parts of partitioned states differ from domestic tourists because the individual must cross a border and probably use a different currency. However, they are not quite international tourists either because the border being crossed is not generally recognized, and the political controls may not be the same for partitioned-state tourists as for tourists from other countries (Butler and Mao 1996: 28).

In addition to acting as a barrier to tourists, especially those from the other units of the partitioned states, these types of *de facto* boundaries create problems related to data gathering. For example, official WTO statistics for tourism in Cyprus include only those tourists who visit the officially recognized Republic of Cyprus (south), and only tourists who visit South Korea are recorded by the WTO and other information agencies.

Restrictions by home countries

Many governments have established travel restrictions on their own citizens for a variety of reasons. Table 2.1 highlights several of these and host-country imposed restrictions. Documentation is one significant impediment to international travel. Although passports and exit visas are issued routinely for residents of some countries, these documentation requirements can act as impediments when they are disapproved arbitrarily, or when excessive fees

Table 2.1 Host and home country restrictions on travel

Restrictions by host country
- currency limitations (import and export)
- entry visas, limitations on duration of stay
- limitations on where tourists are permitted to travel
- limitations on tourist dealings with local residents
- restrictions on the entry of motor vehicles and boats
- formalities concerning car insurance and drivers licenses
- restrictions on the acquisition of holiday properties
- taxes on foreign tourists

Restrictions by home country
- currency restrictions imposed upon residents
- arduous conditions and procedures for acquiring travel documents
- restrictions on overseas travel
- customs allowances for returning residents
- exit taxes for residents

Source: Ascher 1984; Timothy 1995a; 1998c

are charged (Ascher 1984: 4). Prior to the collapse of communism in Eastern Europe, residents of the east faced tremendous obstacles when traveling to the west. For both ideological and economic reasons, many barriers were erected by the socialist governments of Eastern Europe to deter citizens of the region from traveling outside their own countries, especially to the West. Most of the eastern, socialist countries did allow limited numbers of their citizens to travel abroad, but the process was difficult by all accounts. Potential travelers always required an exit visa. This usually meant securing a permit from their place of employment, a police background check, and if male, proof that their military records would be left safely with officials (Hall 1991a). Some countries required a written invitation from an institution or relative in a non-communist country wherein the inviting party agreed to pay the foreign currency costs of the trip. Exit visas to western countries were much more difficult to secure than visas to other socialist-bloc countries, and travel to other socialist countries via western countries was frowned upon and difficult, even if the route was shorter and more direct (Hall 1991a). Furthermore, passports were valid for only one trip and travelers were required to surrender them back to authorities upon their arrival home.

Another, perhaps even more difficult obstacle for residents of many countries is the issue of currency restrictions, or exchange controls. In the mid-1980s, more than 100 countries imposed restrictions limiting the amount of foreign and national currency their residents were permitted to take abroad (Ascher 1984). This situation has recently changed, however, especially with the transition of Eastern European monies into convertible currencies that can be exchanged at home and abroad. Residents of communist Eastern Europe were not permitted to carry their national currencies abroad, and communist currencies were not exchangeable on the international market. As a result, travelers were required to secure Western money before they could travel abroad, and in many cases their trips had to be purchased with hard currency. This was especially problematic in some countries where, for many years, it was illegal even to possess foreign currency, and in most such nations, it was at best difficult to acquire. Most Eastern European countries even added severe restrictions on the amounts of foreign currencies that could accompany their citizens abroad. For example, in the case of Bulgaria, each traveler was permitted to take only US$20 out of the country up until recent political reforms (Carter 1991). Currency restrictions limit expenditures on goods and services outside the traveler's home country, and as Ascher (1984) pointed out, a very low allowance could impede outbound travel altogether. Exchange limits imposed on nationals in their travels abroad are generally put into place in an effort to keep scarce foreign exchange in the country, to improve the country's balance of payments situation, and to discourage citizens from traveling abroad. A related barrier, in some cases huge fees are levied on citizens to discourage international travel, thereby decreasing the leakage of valuable foreign exchange outside the country, and to provide a monitoring system against citizens. For example, a one million Rupiah (US$132 in October 1998) exit fee for residents of Indonesia, together with the high cost of passports and foreign currency, prevents many Indonesians from traveling abroad.

Restrictions by the host country

Many of the restrictions placed on incoming tourists will be discussed later in this chapter in the section on perceived barriers. However, a few items deserve mention here (see Table 2.1). Some countries refuse to issue tourist visas to citizens of certain countries. This is clearly a real barrier since travelers are not permitted to enter the host country without a visa. For example, North Korea, for many years, refused to issue visas to Japanese, American, South African, and Israeli tourists (Hall 1986b), although this policy has been loosened in recent years. Some countries' policies of only issuing group visas curtails the development of individual tourism, which is one of the fastest growing sectors of global tourism. Other governments place stricter requirements on individual visas than they do on group visas. For example, Uzbekistan, as of 1996, still required visas for entry into the country. Group visas were relatively easy to obtain, but individual visas required an official letter of invitation from within the country (Airey and Shackley 1997).

For tourism to flourish, tourists require freedom of access and movement. In centrally-planned economies, attitudes range from severe restrictions on tourist movement (e.g. Albania before the 1990s and North Korea) to limited access to certain areas (e.g. the former German Democratic Republic) to almost complete freedom (e.g. the former Yugoslavia) (Buckley and Witt

1990: 16). The fact that foreigners are still not permitted to travel freely in certain countries creates additional barriers to incoming traffic. North Korea, for example, only accepts tourists in groups of ten or more, and their itineraries are fixed, interaction with local residents being strictly curtailed (Hall 1984; 1990b).

Demarcation methods and fortifications

In most cases, where negative relations exist between adjacent countries, boundaries tend to be clearly marked, and in many cases heavily fortified (Plate 2.1). As part of the objective, or phenomenal, environment, a heavily fortified political boundary clearly affects the flow of information between groups of people and prevents the actual crossing of the border, especially along borders where no, or few, crossings exist. Naturally this situation creates a significant impediment to tourism.

At its inception in 1949, the German Democratic Republic (East Germany) began to control the movement of people into and out of its territory. In the early 1960s, territorial demarcation became especially important to the communist regime, so that East Germany spent approximately US$2.5 million a kilometer to construct a 1,393-km impenetrable boundary with its western neighbor (Ritter and Hajdu 1989). This boundary between west and east consisted of metal and cement fences, deep anti-vehicular

Plate 2.1 The Israel–Lebanon border is an example of a heavily fortified boundary that represents a physical barrier to tourism
Source: Reprinted with permission of Chad F. Emmett

trenches, minefields, guard dogs, observation towers, floodlights, and co crete walls (Buchholz 1994; Ritter and Hajdu 1989). Even after the Ber Wall came down, and East Berlin residents were permitted to cross freely to the western part of the city, East Germans outside the city were still separated from the west by kilometers of barbed wire and landmines (Reaves 1989).

Similarly, as part of the fortification of the USSR–Finland border, a frontier zone was established on the Finland side that ranged from one to three kilometers in width. Non-residents of the frontier zone in Finland were not permitted to enter without a special permit and photographing anything inside the zone was also prohibited. In order to secure peace along the frontier and to maintain personal safety, the Ministry of the Interior (1974) of Finland issued a statement of appropriate behavior within the frontier zone. The minimum penalty for disobedience to these border-zone rules was a fine or imprisonment. The rules included:

- no crossing of the frontier without permission;
- no moving within a strip of land comprising an area of 4–5 meters width along the border;
- no photographing of the frontier or the borderland of the neighboring country;
- no shouting over the frontier or to speak to the frontier authorities or civilians of the neighboring country;
- no throwing objects, documents, newspaper, or anything similar over the frontier;
- no casting light upon the frontier or any area of the neighboring country;
- no careless use of fire, so that it does not cross over the frontier;
- no other behavior, which disturbs the frontier peace, and conduct contrary to good manners and customs in the vicinity of the frontier.

Political conflicts

Research has shown that violence and political instability can have grave effects on tourism (Hall 1994b; Sönmez 1998). Roehl's (1995) examination of the aftermath of the 1989 Tiananmen Square incident found that the event severely reduced foreign tourist arrivals in China for a couple of years because it interrupted growth from a number of key markets, such as Japan, the United States, and Western Europe. Border regions have always been plagued with disagreements, disputes, and political sensitivity – problems that can indeed function as major barriers to tourism. Tourism, in common with many other services industries, is delicate. People want to feel safe while traveling abroad, and if their safety is in question, fewer people will travel, or they will choose alternative, safer destinations. Wars, border disputes, *coups d'etat*, and crime all contribute to unsafe conditions and usually result in sheer declines in tourist flows and border closings (*Nezavisimaya*

Gazeta 1994; *Sevodnya* 1994). This precludes the successful development of tourism when a destination's image is shattered and when its natural and cultural resources are destroyed, either as a byproduct of war or as deliberate targets (Bumbaru 1992).

Wars

The purpose of this section is not to discuss the effects of war and political unrest specifically on tourism, for this has already been done (see Hall 1994b; Sönmez 1998). Rather, the objective is to discuss the spill-over effects of war and unrest on tourism across political boundaries and how they can actually create impassable barriers for tourism. It is clear that wars influence the flows of tourism to the countries that are directly involved in the conflict. However, war and its resultant negative image have no respect for international boundaries. Even when conflicts are entirely contained within one or two countries, the effects are much more far-reaching. For example, on a global scale, the Gulf War of 1991 was blamed for a sluggish growth of international arrivals in places as far away as Southeast Asia and Australia. Travelers generally feared for their safety in all parts of the world, which affected international travel patterns notably in 1991 and 1992. National plans for Visit Indonesia Year 1991 were frustrated by the onset of the Gulf War. Knowledge that Indonesia is predominantly a Muslim country added to the reluctance of Western tourists to travel there (Borsuk 1991).

On a smaller scale, conflicts in one country often negatively influence tourism in contiguous countries. Richter (1992) pointed out that political instability in one country may affect tourism in another country, even when the contention does not actually spill across the border. Disturbances in one nation can make entire regions appear unstable. For example, Croatia continued to feel the negative effects of the ongoing war in neighboring Bosnia Herzegovina long after its own war had ended. According to Panic-Kombol (1996: 21), the war in Bosnia has generated deep negative connotations among potential tourists that Croatia is a dangerous tourist destination. Local analysts fear that this could hinder Croatia in achieving its former level of tourism. Similarly, Bachvarov (1979) reported that the Soviet invasion of Czechoslovakia in the 1960s, the Turkish invasion of Cyprus in 1974, the civil war in Lebanon, and the Arab–Israeli wars have all had a negative impact on international tourism to the Balkans region as well. According to Mansfeld (1996: 267), 'This phenomenon of innocent countries that suffer from other countries' disputes is quite common. Thus, Indian and Maldives tourism have suffered from Sri Lankan terrorism, Pakistani tourism was negatively affected by the war in Afghanistan and the Uganda *coups* deterred East African tourism.' Mansfeld (1996) asserts that this is in large part a result of the role of the media in covering these events. Much of what journalists report is exaggerated and fails to differentiate international problems from national ones.

Surprisingly, not all effects of war on neighboring countries are negative. According to Mihalič (1996), the Yugoslavian wars of the early 1990s increased Austria's tourist receipts by redirecting Yugoslavia's market demand to Austria. Similarly, once peace had been established in Slovenia following its independence from Yugoslavia, Slovenian outgoing tourism turned to Austria instead of Croatia. Mansfeld (1996) suggests that the Arab–Israeli wars obviously have caused a decline in tourism in Israel and its Arab neighbors. However, other countries in the eastern Mediterranean may have gained from these conflicts, as tourists preferred to visit the 'quiet' side of the region, such as Greece or Turkey.

As mentioned earlier, wars can also create political boundaries, which usually function as real barriers to tourism. In many parts of the world, ceasefire lines have become in effect international boundary lines. For example, although the Korean War has never officially ended, a ceasefire line was established, bisecting the Korean Peninsula into two political units in 1953. This border, known as the Demilitarized Zone (DMZ), was established along the ceasefire line between the northern and southern portions of Korea. Other borders have been forced upon weak nations by powerful neighbors. China's northern boundary with Russia was effectively determined by a powerful Russia in the mid-1800s when China was beleaguered by internal problems, which rendered it unable to resist (Prescott 1987).

Prior to the partition of Cyprus in 1974, tourism had reached a significant level of prosperity and was one of the country's major sources of foreign exchange. Approximately 20,000 tourists arrived in Cyprus in 1960. That number had grown to nearly 140,000 in 1970, and by 1973 foreign tourist arrivals had doubled to nearly 300,000. The island's two main resorts, Famagusta and Kyrenia, alone contained 65 percent of the island's hotel bed capacity in 1973 and hosted 63 percent of the island's inbound tourists the same year (Andronicou 1979; Witt 1991).

This success was interrupted in July 1974 when Turkish military forces entered the island and many lives were lost at the hands of both the Turks and the Greeks. The intervention was sparked by a *coup* against the country's leader, organized by extremists who desired unification with Greece. Turkey and Cyprus' Turkish population feared that the *coup* would lead to a rapid annexation of Cyprus by Greece. After negotiations failed to produce the desired outcomes, the Turkish government landed military forces on the island and occupied the northern third of the land mass. Once the occupation was complete, the Turkish military declared a ceasefire line in August 1974 (Kliot and Mansfeld 1997). Between July 1974 and December 1975, approximately 185,000 Greek Cypriots moved southward and about 45,000 Turkish Cypriots migrated north into the occupied zone (Grundy-Warr 1994). When the island-wide ceasefire was declared, the political landscape of Cyprus was altered dramatically. 'An artificial line cut through the island like a cheese-wire, cutting off villages from their fields, splitting streams and underground water resources, truncating roads and power-lines' (Grundy-Warr 1994: 79). The UN peacekeeping troops then had the arduous task of demarcating a buffer zone between the Turkish ceasefire line and the forward front of the Greek Cypriot troops. This demilitarized zone extended 180 kilometers from Kokkina in the northwest to the southeast coast just south of Famagusta (Figure 2.2).

This barrier, known locally as the Green Line or Attila Line, has become a major impediment to tourism in Cyprus. The best beaches and the two most prominent resort communities, Kyrenia and Famagusta, 82 percent of the tourist accommodation, 96 percent of the hotels under construction, over 60 percent of the groundwater supply, the island's best agricultural land, and Nicosia's international airport were cut off from residents of the south and the main flow of tourists by the *de facto* boundary (Andronicou 1979; Kammas 1991).

Immediately following the political upheaval and division of the island, tourism declined sharply (Andronicou 1979; Ioannides 1992; Witt 1991). Tourist arrivals measuring nearly 300,000 in 1973 immediately dropped to 150,000 in 1974 and 47,100 in 1975 as a result of the conflict, but this downward trend lasted only a couple of years. Since the events of 1974, the Republic of Cyprus (the Greek south and the only government recognized by the international community) has experienced a remarkable recovery with regard to its tourism industry. Several new resorts have been developed in

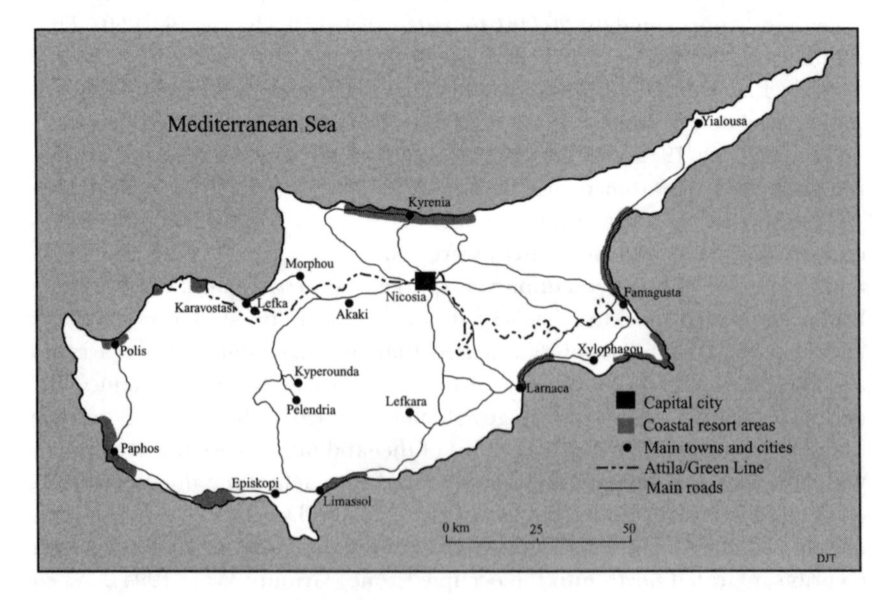

Figure 2.2 The divided island of Cyprus

attractive locations and annual tourist numbers regularly reach
millions. In contrast, tourism in the north, known locally and in T
the Turkish Republic of Northern Cyprus, has only recently startec
signs of significant growth in tourism (Lockhart 1993; 1997b; Lockhart and Ashton 1990; Sönmez and Apostolopoulos 2000). This delayed development is largely a result of international economic sanctions and boundary restrictions, which seriously limited tourist flows and foreign investment (Scott 1995). In addition, Turkish authorities emphasized the development of agriculture, textile milling, and food processing as means for economic development in the years following the division, at the expense of developing tourism.

Wars themselves can also act as real political barriers to tourism development in the sense that tourism services, infrastructure, and attractions are sometimes targeted, destroyed, and looted by rivaling parties. In many countries, public funds, which might be used to develop tourism, are depleted on war efforts. This can have long-term effects for it often leads to the neglect of damaged resources, such as historic sites and tourism-related infrastructure. Furthermore, wars make some tourist attractions physically inaccessible, and barricades are commonly set up around zones of conflict to monitor and deter would-be visitors to the area.

International border and territorial disputes

Borders and territory have been the source of countless international conflicts throughout history. There are few places in the world today that have not experienced border altercations at one time or another. In fact, many countries are still involved in heated debates with adjoining states over such things as the exact location of their common frontier, one country's autocratic manipulation of the boundary and its functions, and the misuse, or illegal consumption, of shared resources, such as fish, water, and oil. For example, even with their best-friends reputation, Canada and the United States have experienced difficulties in recent years along their western border pertaining to fishing rights. Officials in Ottawa accuse American fishers of infringing on Canadian sovereignty by fishing in Canadian territorial waters without permission, and Canadian boats have been seized by US officials for reportedly fishing in American waters (Vesilind 1990).

Russia and China are still involved in a dispute over the exact location of the international border line within in the Amur River. At one point a substantial island is located in the river and if a binding decision was made regarding sovereignty over the island, one of the countries would stand to lose over 332 square kilometers of territory. The tragic war between Iran and Iraq during the 1980s was based largely on Iraq's despotic manipulation of the international border in the Shatt-al-Arab in order to restrict western Iran's access to river transportation. Many more examples of such conflicts abound (see Allcock *et al.* 1992; Lee 1980).

Several border disputes directly involve important tourism resources and destinations. For well over a century, Thailand and Cambodia have been involved in a major territorial dispute. Preah Vihear is an ancient temple complex located near the border of the two countries. It is one of the best examples of Khmer architecture and one of the most impressive temples in Southeast Asia, possessing great tourism potential. The problem lies in the fact that its location has been strongly contested between the two countries, and each has at some point controlled ownership of it. Several disputes have ensued regarding where the border lies, and the temple has changed hands several times between the Thais and Cambodians. Finally, in 1962, the location of the border was fixed by a decision of the International Court of Justice. This resolution confirmed that the temple was indeed located just inside Cambodia (Leifer 1962; Singh 1962; St John 1994).

Between 1962 and 1992, the Thai–Cambodia borderlands were inundated with violence by communist guerillas on both sides. Despite the region's significant tourism potential, this violence frustrated the development of tourism during that period. However, after successful negotiations between the Cambodian and Thai governments in the early 1990s, and a Khmer Rouge ceasefire in Cambodia, the ruins of Preah Vihear were finally opened to tourists early in 1992, with bright hopes for future growth. The site was generally inaccessible from the Cambodian side, so most tourists visited the temple on day trips from Thailand, although arrangements were made so that both countries benefited economically (Cummings 1992). Unfortunately, in July 1993, the Khmer communist rebels reoccupied the temple and temporarily halted tourist visits (St John 1994). The Cambodia–Thailand borderland continues to be a sensitive area. Tourism and the temple of Preah Vihear continue to be pivotal concerns in the border disputes and political unrest in that region, and access by tourists and their safety is nearly always in question.

Another example of international boundary disputes that directly involve tourism is the Taba case between Israel and Egypt. This squabble was a result of the 1979 Camp David peace treaty between the two countries, which mandated the withdrawal of Israel from the Sinai Peninsula, which it seized from Egypt in the Middle East Wars of 1967. The agreement ruled that 'Israel will withdraw all its armed forces and civilians from the Sinai behind the international boundary between Egypt and mandated Palestine ...' and that 'The permanent boundary between Egypt and Israel is the recognized international boundary between Egypt and the former mandated territory of Palestine ...' (Lapidoth 1986: 35). Israel withdrew from the Sinai in 1982, but the two countries did not agree on the exact position of the boundary where it meets the Gulf of Aqaba. Egypt argued that the border should be where it was before the Israeli occupation of 1967, which, according to a 1915 survey map, placed all of Taba in Egypt (see Figure 2.3). Israel, on the other hand, contended that the border depicted on a 1906 survey map, which showed most of Taba in Palestine, was the definitive boundary line (Allcock *et al.* 1992; Drysdale, 1991; Hazan, 1988; Kemp and Ben-Eliezer 2000). What makes this situation unique is that the land area under dispute is simply comprised of a narrow 700-meter long strip of beachfront property. Wayne (1988) indicated that the territory had no real strategic value for either state. However, according to some observers, for Israel it meant adding nearly one kilometer of territory to its Gulf of Aqaba coastline, 'a relatively big slice compared to a coast that does not exceed five miles in all' (Raafat 1983: 18).

Of particular significance in this discussion is the fact that within the small territory of Taba, a premier tourist resort had been developed by the Israelis in 1982. Since Israel considered Taba part of its own territory and felt that it would someday prevail in this claim, the construction of tourist facilities in the disputed area was permitted and even encouraged. The resort included a first-class hotel, a vacation village, and a public beach (Lapidoth 1986). Israel feared that if the arbitration panel decided in favor of Egypt it would lose control of the resort facilities (Kemp and Ben-Eliezer 2000); the world feared that a vote in favor of Egypt would result in increased tension between the two countries (Moffett 1988a; 1988b). Nevertheless, on September 29, 1988, the international arbitration committee ruled in favor of Egypt and ordered the return of Taba, including the two Israeli-built tourist facilities (Friedman 1988; Kemp and Ben-Eliezer 2000). The dictum

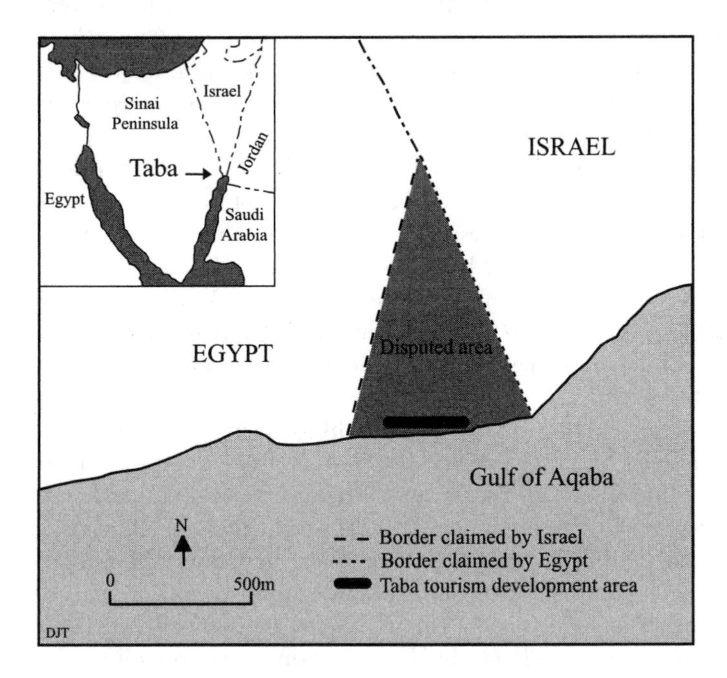

Figure 2.3 The Taba dispute

was binding and Israel agreed to abide by the decision (Hazan 1988), although arrangements were made to negotiate Israeli access to the resort and monetary compensation from Egypt for the resort development. On February 21, 1989, a price for the resort complex was agreed to, and on March 15, 1989, Israel handed control of the small resort territory over to Egypt (*Christian Science Monitor* 1989; *Economist* 1989).

One of the better known border problems in the context of tourism is that between Spain and Gibraltar. Gibraltar, which has been ruled by Britain since 1713, has been persistently disputed by Spain since the mid-1960s. As a result, Spain closed the border and severed all other direct forms of communication with the colony in 1969 (Allcock *et al.* 1992). The border was finally re-opened in 1985 after the UK government agreed to discuss the issue of sovereignty of Gibraltar with Spain. During the period of 1969 to 1985, tourism in southern Europe grew rapidly and the region became one of the most popular destinations for northern Europeans. Prior to 1969, the Gibraltar airport was one of the most significant gateways for British tourists traveling to the Costa del Sol in Spain and a major day-trip destination to which many people traveled by car, bicycle, and foot. However, with the closing of the border, the land-crossing options to and from Spain disappeared. Tourism development in the colony suffered significantly, although it did continue to develop slowly during the years of isolation, mostly from tourists arriving by air or sea, via Morocco. Tourist arrivals in Gibraltar in 1968 were significant at 306,010. In 1970 the number plunged to 140,669; however, numbers soon stabilized at a consistent 120,000–150,000 per year. After the border was opened, over two million tourists arrived in Gibraltar in 1985, 94 percent of which arrived by land, and tourism has continued to thrive (Seekings 1993).

Dozens of other examples exist where border conflicts, some more heated than others, are ongoing and have prevented the growth of tourism in regions which would otherwise be highly desirable tourist destinations. Some of the more obvious examples include the high Karakoram and Himalayan region of India/Pakistan and India/China, the pristine tropical rain forests of Venezuela and Guyana, the isolated and subarctic Patagonian region of Argentina and Chile, and the rugged Caucus Mountains in Armenia, Azerbaijan, and Georgia.

In addition to regular border altercations, fighting over territory not connected to border zones has also been an obstacle to tourism in many parts of the world. This is important to discuss as well because territorial battles are usually fought in an effort to attenuate national sovereignty over foreign or unclaimed territory, in effect extending the political boundaries of a state's realm of control. For example, the invasion of the British-controlled Falkland Islands by Argentina in 1982 was the culmination of tensions resulting from claims of ownership of the islands by both Britain and Argentina. The Argentineans eventually withdrew from the islands but the dispute still has not been resolved. Tourism in the Falklands was seriously affected by this war (Boyd 1997). Many other contemporary examples of territorial disputes exist, such as the tensions that are still strong between Morocco and Algeria over ownership of Western Sahara and the claim by Venezuela that it possesses the legal right to sovereignty over more than half the territory of adjacent Guyana (Prescott 1987).

A further perspective on territorial disputes is that of tourism being used in some regions to support territorial claims (Allcock *et al.* 1992; Hall 1994b). The Taba case described above is a good example of one country's efforts to develop tourism rapidly to assist in upholding territorial claims. Another example is the Spratly Islands. These are a group of some 230 small islands and rocky outcrops located in the South China Sea, comprising merely five square kilometers of land. Only about 25 of the islets are substantial, and few, if any, are viewed as being capable of supporting any kind of civilization. The remaining outcrops are nothing more than rocks, reefs, and uninhabitable atolls. Despite their small land area and seemingly useless existence, the Spratlys are one of the most contested territories in the world today. For strategic, economic, and nationalistic reasons, six countries – China, Vietnam, Taiwan, the Philippines, Brunei, and Malaysia – claim all or part of the Spratlys and their adjoining maritime zones. In 1993 more than 15 percent of the world's international trade passed through the Spratly region (Dobson and Fravel 1997). The area is believed to be rich in oil deposits and mineral resources. Vietnam occupies 21 rocky outcrops, China eight, the Philippines eight, Malaysia four, and Taiwan one (Dobson and Fravel 1997; Walsh 1992). Brunei, although it claims territorial rights, does not occupy any of the islets. Sovereignty over the islands is intensely contested and violent exchanges between national military forces is not uncommon in the region. The entire problem has been complicated further by the 1982 United Nations Law of the Sea, which was drafted to clarify maritime jurisdictions. The law guarantees fishing and mineral rights within a 370 kilometer exclusive economic zone (EEZ) around all national territory, hence the race for new territory. The law also stipulates that EEZs can be established only if the territory in question can support human habitation. Malaysia has been inventive in this respect by developing tourism on one of the islands it claims. Beginning in 1983, this 500-meter long rock received a facelift geared towards tourism. By 1992 it boasted a strong military defense infrastructure, as well as 17 guest houses and some trees anchored in soil brought from the mainland – an effort that cost the Malaysian government over US$30 million (Allcock *et al.* 1992; Tyler 1992). As a result of increased regional tension, however, the project has since been abandoned because tourists' safety could not be guaranteed (Hall 1994b).

For obvious reasons, frontier and territorial disputes necessarily prevent the successful and continuous development of tourism in disturbed borderlands.

These kinds of disputes also hinder the development of tourism within the interior of participating countries, since borders become essentially impenetrable in both real and perceptual terms, so that overland tourist travel between countries becomes impossible or, at best, risky.

Surprisingly, not all boundary and territorial friction results in negative effects on tourism, however. In rare situations, border disputes have actually been a unique attraction for tourists. According to the owner of the resort complex in the disputed Taba territory, one of the attractions of the hotel was the fact that guests could see four countries from their balconies. He further concluded that tourists were drawn to Taba in part 'because we are an international problem. How many people get to stay at an international problem?' (quoted in Drysdale 1991: 205).

Subnational and other administrative border disputes

Border and territorial disputes are not limited to international frontiers. Many subnational polities throughout the world have experienced border-related conflicts with adjoining states, provinces, counties, and municipalities. Earlier this century, interstate boundary disputes were quite common in the United States, for example. During the 1920s Texas waged a legal war with its neighbors, Oklahoma (Billington 1959; Carpenter 1925) and Louisiana (Andrew 1949), over the exact location of its boundaries, claiming territory on the opposite banks of the Red and Sabine Rivers. Texas and New Mexico also battled over their common border in the Rio Grande (Bowden 1959). Luckily most of these disputes have been settled and do not have a significant bearing on tourism. Although the core of these arguments was not tourism, such subnational border conflicts may have significant effects on tourism. Many of the regions of North America that have experienced border altercations, such as the Texas cases above, Wisconsin–Michigan (Martin 1938), Georgia–South Carolina (de Vorsey 1982), and Utah–Arizona (Comeaux 1982) are regions where tourism is an important economic activity. Results of territorial disputes mean the difference between which state can develop tourism resources and infrastructure in a given area, as well as receive the economic benefits of tourist activities.

A centuries-old border dispute between New Hampshire and Massachusetts (USA) prevented full cooperative efforts between the two states in terms of trade and tourism promotion until the battle was declared officially over in 1991. Once the declaration was official, the two states agreed to meet in an effort to coordinate interstate trade and tourism development (Biddle 1991). In a similar situation, the states of New York and New Jersey have been battling over control of Ellis Island in New York harbor for many years. The island is a highly significant heritage tourist attraction owing to its former days as the processing center for immigrants to the United States in the early 1900s. Based on an original 1834 agreement, the US Supreme Court ruled on May 26, 1998, that the original 1.3 hectares of the island belong to New York, while the remaining 9.8 hectares, created by landfilling during the late 1800s and early 1900s, are under the jurisdiction of New Jersey. The buildings on Ellis Island are owned and operated by the national government as a heritage site, but the new ruling has divided some of the buildings between the two states. Tourists are unlikely to notice any differences in jurisdictional changes, but now that the dispute has been resolved, ownership of sales tax revenue can be cleared up, and possible future development plans can be made by each state or in cooperation (Greenhouse 1998: 21).

Border disputes between even lower-order jurisdictions, such as counties and municipalities have been, and continue to be, problems in many regions (Hasson and Razin 1990; Sonderegger 1996; Woolley 1996; Yiftachel 1990). While not generally as severe as international conflicts, contentions such as these on a lower political hierarchy still can have major implications for the development of tourism, especially in terms of taxation and promotion.

In addition to international and subnational frontiers, the borders of national parks and other special administrative units commonly undergo changes as well. In the name of conservation, park officials and environmental activists fight to extend park boundaries. Efforts to maintain and expand the limits of parklands frequently face uncompromising challenges. Among others, these challenges include hunting, intensive agriculture, small- and large-scale forestry, mining, oil and natural gas drilling, private land ownership, urban encroachment, and industrial pollution. Border changes are often at the center of heated disputes between government agencies, between local residents and park officials, and between business owners and park administrators.

In the United States, Yellowstone National Park is facing harsh predictions about the long-term ecological effects of a large gold-mining project planned for construction just outside the park's boundaries in the state of Montana (Beers 1995). Similar warnings were issued and disputes broke out in the late 1980s between environmental conservationists and business and government groups in the Canadian province of Alberta, where a major oil company was planning to drill extensively for natural gas just outside Waterton Lakes National Park (van Tighem 1988). Dozens of additional accounts have been recorded of the environmental effects on parklands of agriculture, urban sprawl, logging, and industrial development located near national park boundaries throughout the world (Barker and Miller 1995; Kenney 1991).

Furthermore, efforts by the US National Park Service (USNPS) to expand park boundaries have been met by a great deal of opposition by local residents in various parts of the country. In the case of Grand Canyon National Park, Arizona, officials and other conservationists have fought to expand the park's territory, so that more of the surrounding wilderness might be protected under federal law. One of the project's primary obstacles is that only 60 percent of the adjacent territory is federally owned. Native Americans own 25 percent, private citizens own nine percent, and six percent belongs to the state. Some of the most passionate opposition to parkland expansion has come from the Havasupai Native Americans, who would lose part of their reservation lands if the park's boundaries were expanded (Morehouse 1996).

Similarly, in the eastern United States, many residents have bitterly opposed the expansion of Shenandoah National Park, Virginia, for many years. Residents who live around the park feel that they have lost control over their own property and are being forced to comply with rules set by intrusive outsiders. In fact, outright hostility has developed on the part of local inhabitants towards the USNPS (Fordney 1996). This ill will has driven many residents to organize networks of property-rights groups to lobby actively against park authorities and fight the push by USNPS to expand the park. Shenandoah National Park has grown by 870 hectares since 1980 through land donations and exchanges. In response, the property-rights groups recently undertook legislative action to prevent this type of territorial accretion from continuing. If the proposed bill is passed, it will freeze the borders of Shenandoah National Park to land it now owns and would require congressional approval before land could be donated to the park (Fordney 1996).

Coups d'etat

Teye (1988) has provided a detailed explanation of the effects of *coups d'etat* on tourism. Generally speaking, these military takeovers have significant effects on tourist arrivals and perceptions of disturbed destinations abroad. However, *coups* have particular relevance to this discussion of political boundaries and tourism, particularly since one of the most immediate actions taken by successful *coup* leaders is to seal all land borders, as well as air and sea gateways into and out of the country. These actions of isolation can last days, weeks, or even years. Although the main purposes of such actions are to keep 'wanted' people from escaping and to prevent reinforcements for rebels from entering the country, the closing of borders and other gateways clearly has significant impacts on international tourism (Teye 1988).

Border closures mean that tourists already in transit are not permitted to enter the country, and visitors already in the country become stranded when they are not permitted to leave. The closing of outside communications links often accompanies the closure of national borders. When communication to the outside world is severed, a very strenuous and complex situation is created for international visitors (Teye 1988).

Crime

Several authors have recognized the relationships between tourism and crime, and have proposed that criminal activities, especially economic crimes such as theft, pick pocketing, and drug dealing, are sometimes directly related to the growth of tourism (e.g. Jud 1975; Mathieson and Wall 1982; Pizam 1982; Ryan 1993). Understandably, large population centers and regions of high levels of tourism appear to be especially prone to such criminal activities.

Border areas are also notorious as prime locations for economic crimes, since they often possess higher numbers of transient residents and travelers that contribute to a greater degree of local restlessness (Lin and Loeb 1977). Furthermore, frontier zones are fitting venues for illegal smuggling and other forms of trans-border organized crime, especially when one government is less restrictive in terms of law enforcement against drugs, smuggling of consumer products, racketeering, and other illicit activities. International boundary locations also provide opportunities for bandits to escape, sometimes with relative ease, into a foreign country.

For many years, Big Bend National Park in Texas, located along the US–Mexico border, has experienced problems of shootings and automobile break-ins. Trail-head parking lots tend to be a favorite location for thieves. According to one park ranger, 'It's very easy for someone to ride across the river on a horse, hit a vehicle, and go back across. You'd be surprised at what they take ... water jugs, gasoline, tires, [and] kerosene lanterns' (*Outside* 1994: 63). Within the park dozens of automobile break-ins are reported every year, and in 1993 over US$5 million worth of drugs were seized. Within the national park, on the US side of the river, a group of 30 mounted and armed Mexican men hangs out, and for $50, provides an informal towing and ferrying service for people and vehicles crossing in either direction, no questions asked (*Outside* 1994: 66).

In some places law enforcers are more permissive on one side of a frontier than they are on the other side as regards illegal activities and as a result, these activities often spill over into neighboring border regions. The frontier between Iran and Azerbaijan was opened in 1991 with the collapse of the Soviet Union. Since then the border has become a popular location for drug trafficking, smuggling, and racketeering. Gangs have formed in the region and gained control of customs agents and other law enforcement officers (*World Press Review* 1993). Crime, like wars and border skirmishes, can create fear within tourists which will cause them to choose alternative destinations. If a borderland destination's reputation becomes tainted with tales of criminal activities, the local tourism industry can be damaged far into the future.

Perceived barriers

While not all borders present real or physical barriers to tourism, they can act as significant perceived, or functional, barriers. Even borders between friendly neighbors can create perceived barriers, as the French–Belgian border did in the 1970s (Dewailly 1979), and as the US–Canada border continues to do (McKenna 1997a; Slowe 1994). The degree to which this is so depends on several factors including the political relationships between adjacent nations, the range of socio-cultural differences on opposite sides of a border, economic conditions, and the perceptions of potential border crossers regarding what lies on the other side. Manifestations of these issues include border formalities, cultural differences, administrative differences, and costs. After a brief look at the possible psychological reasons for perceived barriers, these issues are examined in the sections that follow.

Budd (1990) suggested that fear is the source of many people's reluctance to cross borders. In a sense, the fear associated with traversing political boundaries, whether a result of strict regulations, ignorance, or a fear of cultural differences on the opposite side, is rather like a person's fear of flying, often created by a sense of danger or loss of personal control (Aronson 1971; Marks 1987). At one end of the spectrum crossing a border may create a sense of risk and adventure, while at the other end outright fear with its accompanying physiological symptoms: rapid breathing, nausea, trembling, increased irritability, and headaches (Aronson 1971). Fear is created in a variety of ways. It may be created through direct life experiences (Campos *et al.* 1992) or through social conditioning and reinforcement. Such fears and phobias commonly prevent people from participating in certain otherwise desirable activities (Pearce *et al.* 1996; Tavris and Wade 1995).

Some people's fears, if not better described as agoraphobia or social phobia, fit into the category of specific phobias (Thorpe and Olson 1997). A few types of specific fears help explain why borders often function as perceptual barriers to travel. Perhaps the best example is neophobia, or the fear of new and unfamiliar places and objects. Many people exhibit neophobic behavior by refraining from crossing into strange and different places. In effect, they would be stepping outside their comfort zone, or zone of control, into some place alien. Leimgruber (1989) wrote that people's perceptions of boundaries are molded by a feeling of going beyond, of crossing into a different political, social, and economic realm. Similarly, Ryden (1993) suggests that borders imply a transition between realms of experience. They are lines between life as lived in one place and life as lived in another. This is precisely what many people fear. A normal reaction to this type of specific fear would be escape or avoidance (Marks 1987), such as some people's behavior in relation to crossing a border.

A related concept is that of objective and subjective environments. Researchers have long realized the need to distinguish between the objective, or real, environment, and an individual's subjective, or mental image of, that environment (Lowenthal 1961; Reynolds and McNulty 1968). For many years, geographers and other social scientists have used real (also known as phenomenal) environments as settings for their research (Reynolds and

McNulty 1968). However, more recently scholars have begun to realize that the subjective, or perceptual, image of the environment also creates various types of human behavior (Edwards *et al.* 1994; Kirby 1996) and does in fact determine people's action space.

Human behavior is not a simple reaction to the objective, or phenomenal, environment, but a reaction to a partial and distorted psychological representation of that environment. A person's subjective space governs his or her actions and it is in this context that rational human behavior begins and decisions are made (Goodall 1987: 37). The concept of subjective environment borrows from Gestalt psychology to suggest that the objective environment may have very different meanings for people of different cultures (Kirk 1963). The behavioral environment then is a product of individuals or societies interpreting reality from within a particular cultural setting. Socio-economic, familial, and educational status among populations, as well as national origins determine significantly many of the travel choices made by people from certain populations – cultural and national differences may determine what degree of cross-border interaction takes place (Wackermann 1979).

The extent to which a government allows its people to experience the border influences their perceptions of it. People who live a distance from the border will have a different perception of it from people who live in daily contact with it (Leimgruber 1989; Pagnini 1979). This behavior demonstrates a tendency to ignore the actual features of the border (Kirby 1996) in favor of a standardized social perception of what the border is and how it functions. Phenomena, places or events outside the behavioral, or subjective, space have no relevance to, and no influence on, conscious decision making and human behavior (Small and Witherick 1995: 19). Merrett (1991: 23) suggested that 'people who live close to the boundary, but far from a border-crossing point, do not include areas across the boundary among their action spaces.' Leimgruber (1989) supported this concept by stating that boundaries themselves are of little importance in the spatial organization of behavior. Rather it is the way people perceive them that creates spatial patterns in the landscape.

Border formalities and restrictions

Travelers commonly view borders as barriers in the sense that they must present proof of citizenship, declare goods purchased, and respond to a series of questions from intimidating immigration and customs officials (Smith 1984; Timothy 1995a; 1995b) (see Plate 2.2). According to interviews with US and Canadian border officials by the author in 1998, frontier workers are supposed to be intimidating. Many people become nervous crossing borders even when they have nothing to hide. In fact, one US immigration official who had worked on the border for ten years admitted,

Plate 2.2 While this Canadian border station appears unthreatening, having to report for inspection is a hassle and possibly even a fearful experience for some travelers

'Even I get nervous going over there [to Canada] because I have to go through inspection procedures too.' Travelers are typically scrutinized regarding the value of the goods they purchased in the other country, whether or not they are carrying firearms or prohibited food products, their citizenship, and where they live. Frontier formalities are continuously changing throughout the world, however. In some regions, they are becoming stricter; in other areas they are becoming looser – a concept that will be examined further in chapter 5.

As part of recent crackdowns on the crossing of illegal aliens into the United States, that country enacted a new law in 1997 known as the Illegal Immigration Reform and Immigrant Responsibility Act. Section 110 of the new law requires the documentation of every non-US citizen entering and leaving the country. Although geared specifically toward immigration problems along the US–Mexico border, the law in its present form will apply to the nearly 100 million annual trips made by Canadians to the US as well. Considerations are under way to install an automated system that would check travel documents of every person entering and leaving the country, and special lanes would be provided for US citizens and non-US citizens. Canadians fear that this will mean that they will have to acquire US visas, and other observers predict that the new measures will create massive traffic jams stretching for tens of kilometers at the busiest border crossings (Canadian Press 1997a; Griffith 1997; McKenna 1997a). Members of congress from US border states and Canadian diplomats are fighting the new law and are seeking an exemption for Canadians (Beltrame 1997; McKenna 1997b). An action of this nature on the part of the US congress will obviously result in fewer Canadians and Americans crossing the border for all types of travel. The border, which is already viewed by many as a barrier, will clearly become even more of a barrier if, and when, such legislation is operationalized. Fortunately for citizens of both countries, the bill became law on October 1, 1998, but US officials still lack the know-how and facilities to implement the law, so that it will likely not be put in place until at least 2001 (Canadian Press 1997b; McKenna 1998).

Similarly, in the mid-1980s in a campaign to rid the country of terrorists, France began requiring visas for nationals who had until that time been exempt from visa demands. As a result, the annual rate of growth in tourist numbers to France decreased dramatically because the new restrictions became too much of a hassle for many foreign tourists (Economist Intelligence Unit 1995). The new requirements also included advanced visas for entry into French overseas territories. Some territories, such as Martinique and Guadeloupe, ignored their home government's orders and issued visas on arrival at the airport. Many airlines serving the islands, out of respect for France's policy, required passengers to show visas before boarding the plane. This circumstance created a unique situation on the island of St Maarten. There is no international airport on the French side of the island so travelers land on the Dutch side and take local transportation across the border. There have never been customs and immigration inspections crossing from one side of the island to the other, but since air arrivals are on the Dutch side, French officials considered setting up formal border controls on the island to monitor the visa situation (Sturken 1986), although this never came to fruition. Although France technically required visas for its side of the island, free access between the two sides is assured by international treaty. Instituting these kinds of border restrictions on St Maarten would have no doubt affected tourism to the island.

Cultural differences

While some borders divide different cultural groups, others divide similar social groups. The degree of cultural similarity on both sides of a border will be determined largely by the history of the border and to what degree residents are permitted to interact. Boundaries of recent vintage will be more likely to bisect similar cultural groups. Long-established borders, on the other hand, will have separated societies for a long enough time that each will have developed individually from those across the border. With time, values change and social representations of the world are altered. This is why many societies, which appear to be quite similar, possess different value systems on opposite sides of a border. For example, Canada is much more internationally oriented than the United States when it comes to promoting and developing tourism. The United States on the other hand tends to delegate tourism development authority to the individual states who then often focus more on domestic aspects of development and promotion. Recent policy changes in the US have further decreased the internationalization of tourism administration and promotion (Brewton and Withiam 1998).

When languages and cultures are different on opposite sides of a border, an additional barrier is created. Potential tourists may fear driving into another country if road signs are in a foreign language or if residents do not speak the same language as the tourists. This problem is compounded even more when residents on one side are unaware of the culture on the other side. For example, many Americans are ignorant about what lies on the Canadian side of the border in terms of culture. They understand that Canadians and Americans share many social and cultural similarities and that many Canadians speak French. Some Americans believe that one must speak and understand French to get along in Canada as a tourist – anywhere in Canada. The author's experience interviewing American tourists on the US side of Niagara Falls revealed that several people were wary about crossing the border into Ontario because they did not speak French and were surprised to learn that relatively few residents of Niagara Falls, Ontario, speak French themselves.

Administrative differences

While travel to the west by citizens of Eastern European countries was regulated by real barriers as discussed earlier, travel by westerners to Eastern Europe was hindered more by a perceptual barrier created by complex border formalities, strict currency controls, prepaid accommodations, and limited state-designed itineraries, as well as the fact that ideological differences were often in direct opposition to systems at home. It is clear that this kept many western tourists from traveling to the east, although as will be seen in the next chapter, for some it was, and continues to be in some regions, an actual attraction.

Different systems of measurement (e.g. metric versus old English) can create barriers if potential travelers do not understand kilometers or miles. Road conditions and inadequate signage can also deter people from driving across a border. For example, on the Mexican side of the US–Mexico border travelers are often confused because 'a scarcity of highway signs also discourages motor travel, because many Americans, already troubled by the language barrier, frequently get lost' (Budd 1990: 15). Some travelers also fear spending foreign currency. They might not know the actual value or the correct denominations, which can make them feel vulnerable to being 'ripped off' or otherwise losing money.

Cost

Although prohibitive costs were discussed earlier in this chapter as a real barrier created by some governments for people who have to pay to leave their own countries, the concept merits a quick examination in this context as well, for although much less expensive in most cases, fees are sometimes established for people entering a country, which adds an element of perceptual barrier to the barrier equation.

In 1996 a law was passed to levy a border processing fee for every crossing of the Russian border in both ways by people, vehicles, and freight. This new tax, which is separate from airport fees and customs charges, will, according to Agafonov (1996: 2), 'fleece Russia's citizens and guests traveling in either direction at a fixed rate calculated on the basis of the monthly minimum wage.' Every person crossing the border is charged US$10 (*Wall Street Journal* 1996). More is charged if the person is driving a vehicle, and that rate depends on the vehicle size and number of seats. For business people the amount is not large, but for average citizens it is a significant amount of money and having to pay it will result in fewer border crossings by Russians (Mashinsky 1996), and will likely result in fewer foreigners crossing as well.

Minimum currency requirements, which many countries have instituted, also place a cost burden on tourists (Buckley and Witt 1990: 16). This raises the cost of a vacation to these countries, thereby limiting the number and types of travelers who are able to visit.

Situations where goods and services are more expensive in a neighboring jurisdiction can also create barriers. Unfavorable exchange rates, high prices, and high taxes will act as an obstacle to travel. The favorable US dollar exchange rate of the late 1980s and early 1990s for Canadians has become very unfavorable, with the dwindling of the Canadian dollar from an advantageous rate of Can.$1.15 to US$1.00 to a prohibitive rate of Can.$1.50 to US$1.00 (as of October 1998). This has become a significant border constraint to international travel, and most Canadians who flocked to the border in the 1980s and early 1990s (Timothy and Butler 1995) are now staying home (Stinson and Bourette 1998).

Another border-related constraint imposed on travelers by their home country is customs regulations. These can become deterrents to international travel when they restrict what types and amounts of goods can be brought back into the country. Strict customs regulations and high taxes on imported products can function to discourage excess spending abroad and are commonly established as a form of protectionism for the domestic production of certain goods and services. Furthermore, duties paid on goods purchased abroad are a significant source of revenue for many countries.

Functional distance

The combination of these barrier effects results in what is known as functional distance. Human interaction across political boundaries does not appear to be affected as much by real or physical distance as it is by functional distance. For example,

> Two individuals may reside within a few feet of one another, yet, if their apartment entrances face onto opposite streets, the probability of their confronting one another may be low. Thus while the physical or geographical distance separating them may be slight, the functional distance ... may be great.
>
> (Reynolds and McNulty 1968: 27)

According to Merrett (1991: 23), the fewer the crossing points, the greater barrier the border is perceived to be.

Functional distance may in fact deter more travelers than real distance. According to Smith (1984: 37), this has resulted in 'the volume of travel between adjacent US states and Canadian provinces more closely resembl[ing] travel patterns between two distant regions.' He suggests that the international boundary can be viewed as equivalent to an additional distance a visitor must travel – a distance great enough to decrease the total volume of tourist traffic significantly. This concept was recognized by Reynolds and McNulty in 1968. They posed that 'the functional distance separating ... individuals from the opposite area will probably be greater than the actual highway mileage (or travel time) because of the availability of more adequately perceived and accessible destinations in the area of residence' (1968: 33). Mackay (1958: 1) also recognized this concept early on when he stated that 'we may find, for example, that a mile width of river has the same barrier effect for travel as a land distance of 100 miles ... Perhaps, for many types of human activities Windsor is, in geographical terms, nearer to Vancouver than to Detroit.'

Smith (1984: 38) proposed a method for estimating the distance equivalence of international boundaries. Using data describing patterns of vacation travel from mid-western US states that border Canada and straight-line distances between population centers, Smith (1984) estimated the distance equivalence for Americans traveling to Canada. He found that even though Ohio, which does not share a land boundary with Canada, lies only approximately 100 kilometers from the nearest land crossing, the distance equivalence is 898 kilometers. Similarly, for residents of Indiana, whose nearest corner to the Canadian border is less than 180 kilometers, the distance equivalence is 2,940 kilometers. The functional distance for residents of New York state is particularly interesting at 3,526 kilometers, even though the state is located directly against the Canadian border. While this measurement obviously has some utility for marketers and planners of international destinations, it has one major flaw. Many of the states which are very near the US–Canada border are physically large and their populations are widespread. In the case of Ohio, for example, Cincinnati, a city of nearly 400,000 residents on the southern end of the state, actually lies about 400 kilometers from the nearest land crossing into Canada. Therefore, the disparity between the straight-line distance between Toledo (the Ohio city nearest the Canadian border), and Cincinnati is great. To aggregate all of Ohio into one category of distance equivalence does not portray an accurate picture, especially since it may be that residents who live nearer the international border would be more inclined to cross it than those who live further away as Leimgruber (1989) suggested. Smith's (1984) formula, while useful and informative, might best be utilized when vacation-related data are available on a more local level, such as for cities, counties, or regions, such as Northwest Ohio.

Functional distance can be created between neighboring regions as a result of antagonistic relationships. Until recently, Argentina and Chile were involved in a boundary dispute and old antagonisms which resulted in long-standing prejudices on the part of both countries. Even though most issues have been resolved, and few real barriers exist, jealousies and prejudices are dividing rather than unifying these close neighbors on the periphery of South America. As Caviedes (1994: 137) stated, 'One might visualize these two populations as living with their backs towards each other instead of extending their arms across the boundary line.' Owing to these feelings of hostility between the two governments, and local residents, border residents of both countries have tended to depend more on distant regional centers in their own countries, turning their backs to the international border. For example, residents of Rio Turbio and 28 de Noviembre, two small Argentinian border towns, tend to travel 220 kilometers to Rio Gallegos, Argentina, for their services rather than the 18 kilometers to Puerto Natales on the Chilean side, regardless of an absence of topographic and political barriers to communications and transportation. Cultural and political prejudices appear to be the main obstacles against making this international border a line of integration in a region that otherwise 'looks so physically "unifying"' (Caviedes 1994: 139).

Summary

This chapter has examined political boundaries as barriers to travel and to the development of tourism itself. Borders can act as real barriers to tourism, actually preventing the flow of tourists to, and growth of tourism in, some destinations through unfavorable international relations, restrictions by the home and host countries, the physical barriers related to border demarcation, and political conflicts, including wars, border and territorial disputes, *coups d'etat*, and crime. Unfavorable relations bring about restrictions on travel between countries, which often results in tourists having to travel via a third-party nation to reach their final destination. Arduous processes to secure travel documents, excessive fees, and currency controls are some of the deterrents to travel created by home countries. Host countries also restrict travel by refusing to issue visas to citizens of certain countries and by restricting where and when foreigners can travel within the host country. As well, defensive demarcation methods keep people from physically crossing borders. Political conflicts create a reluctance to travel on the part of tourists, and events such as wars and border conflicts obviously prevent tourism from developing in an area in terms of infrastructure and resources.

Perceived barriers also place roadblocks in the way of tourism. Fear or ignorance of border-crossing formalities, cultural differences, administrative differences, and costs contribute to people's reluctance to cross international boundaries. Even when they have nothing to hide, people are often fearful of passing through frontier formalities. Differences in language and currency also add to the perceived barrier effect of borders, as do foreign driving laws and ideological paradigms that are different from those at home. Excessive costs created through border fees, unfavorable exchange rates, and currency restrictions also contribute to the barrier effects of international frontiers.

All of these barrier effects result in what is known as functional distance. This concept suggests that political boundaries add a perceived distance to certain destinations that generally far exceeds real distance. For example, many people are more inclined to travel great distances within their own countries than they are to travel shorter distances but across an international frontier. In practical terms then, a destination that lies just over a border appears much more distant than a destination within one's own country but located much farther away.

3 Borderlands tourism

> In a subtle and totally subjective way, each side of the border feels different; in the space of a few feet we pass from one geographical entity to another which looks exactly the same but is unique, has a different name, is in many ways a completely separate world from the one we just left. Look: it is even a different color on the map. This sense of passing from one world to another, of encompassing within a few steps two realms of experience, enchants and fascinates
>
> (Ryden 1993: 1)

Introduction

The magnitude of tourism in border regions is immense, and many of the world's most popular attractions are located near borders or right on them. Several authors have described the development of tourism in these unique peripheral settings (e.g. Baerresen 1983; Boyd 1999; Butler 1996; Essex and Gibb 1989; Gibbons and Fish 1987; Herzog 1990; Krakover 1985; Lintner 1991a; Mikus 1986; Minghi 1981; Ruppert 1979; Slowe 1991; Timothy 1995b; 2000a).

Peripheral locations can be viewed from at least two perspectives: in a global politico-economic sense, such as the less-developed countries of the world, and in a regional sense, such as border areas and areas of physical isolation (Timothy and White 1999; Weaver 1998). Because many of today's tourists tend to demand pristine environments and off-the-beaten-path destinations, much of the growth of tourism has occurred in the peripheral and isolated regions of the world. Christaller (1955; 1963) recognized this pattern early on when he suggested that tourism 'avoids central places and the agglomerations of industry' and that 'tourism is drawn to the periphery of settlement districts as it searches for a position on the highest mountain, in the most lonely woods, along the remotest beaches' (1963: 95). In his theory of central places, he explained how economic forces pull towards population centers, 'whereas in the theory of tourism the opposite is true – a tendency towards the periphery, away from the familiar scene towards distant places' (von Böventer 1969: 118). Christaller was referring to physiographic peripheries, but the same holds true for political peripheries in the national context as well.

Several authors have discussed peripherality in the context of tourism in developing regions, pointing out the dependency, or neo-colonial, relationships that exist in many parts of the world as peripheral regions rely on support from the core to survive (Høivik and Heiberg 1980; Keller 1984; Turner and Ash 1975). In a regional sense, however, peripheries can be seen as areas near national borders or regions of difficult climate and topography. Often areas on the geographical periphery are also the poorest sections of countries or regions, as is the case in Europe (Page 1994; Wanhill 1997). As a result, tourism is often targeted as a means for economic development in borderlands and other peripheral locations (Friedmann 1966; Husbands 1981). Tourism also develops in frontier regions because these provide some of the most pristine natural landscapes and engender a mythical frontier image that appeals to tourists (Butler 1996).

Borderlands and their perceived remoteness appeal to the human psyche – something that has been 'harnessed through creative and innovative marketing to boost tourist arrivals' (Page 1994: 45). Planning and promotional efforts in some frontier communities focus on their borderlands locations as a competitive advantage. However, since border areas exert a natural appeal among tourists, tourism sometimes grows spontaneously without a great deal of intentional planning efforts. This is particularly the case where large population bases exist near frontier zones and where policies and practices differ on opposite sides of a border. This chapter is concerned more with the spontaneous growth of borderlands tourism and the appeal that borders exert. Chapter 6 will discuss the planning and deliberate development of borderlands tourism in more depth.

Spatially, tourism in borderlands can be viewed from two primary perspectives: (1) the borderlines themselves as objects of tourist attention, and (2) tourism that does not focus directly on the border itself, but which owes its existence to its relative location near the border. The elements of these perspectives are examined in the sections that follow.

Borderlines

While the border, representing differences in language, culture, and politics, can act as a deterrent to some travelers, it is an attraction for others. Eriksson (1979) suggested that to be close to a foreign country for many people is exciting. This appears to have been the case in the Hong Kong New Territories village of Lok Ma Chau, where farmland and traditional ways of life in China could be viewed across the fence from observation posts (Timothy 1995a). And prior to 1991 in Finland, tourists were particularly intrigued by getting close to the USSR border, looking at it with binoculars, and taking pictures of the land beyond (Paasi 1996). Crossing borders can be a motivation for some people to travel. The border formalities and differences in landscape add to the ambience of the trip and can create a sense of romantic nostalgia. One observer reflected that

> Borders have fascinated me since childhood. As a kid, I used to imagine border landscapes: dark rivers, watchtowers, and unknown lands lying beyond them ... Over the years, as I started travelling, borders have been somewhat demystified, but now again, approaching the Finnish–Russian boundary, I was feeling that boyish excitement, an anticipation of mystery.
>
> (Medvedev 1999: 43)

Likewise, dozens of attractions throughout the world have borders and peace between nations as their primary theme, and many tourist attractions or destination areas are bisected by political boundaries even if their themes do not focus on the border itself. Lines of latitude and longitude are also major attractions in some communities. In fact, several areas rely almost entirely on these linear objects for the basis of their tourism industry. Tourist attractions such as these that relate directly to boundaries are the focus of this section.

Demarcation icons

The way a border is marked and what it encloses on the other side have a great deal to do with the extent of its appeal for tourists and the extent to which it is permitted to attract people. Curiosity seekers have for a long time been intrigued with political boundaries and how they are marked (e.g. Griswold 1939). Features of the border landscape, such as highway welcome signs, flags, and customs buildings, may also be a focus of attention for some tourists. In their own right, these icons are attractions since they mark the interface of differences in language and culture, social and economic systems, and political realms (Timothy 1995b; 1998a). According to Ryden (1993), borders carry a certain mystique and fascination. 'They imply a transition between realms of experience, states of being; they draw an ineffable line between life as lived in one place and life as lived in another' (Ryden 1993: 1). Along the Finland–USSR border, the signs prohibiting entry into the frontier zone were commonly an important stop for tourists. This abrupt line where capitalist ideals succumbed to communism was of notable interest to Finns and foreigners alike (Paasi 1996: 274).

> Many Finns used to visit the Finnish–Soviet frontier, especially in the olden days of a sealed border, to feel the mystique of the place, take photos of the prohibitory sign, or even to step into the restricted border zone seeking to experience a geopolitical thrill – taking a small step

> towards the Other, into the realm of shadows. In the last decade, the border became a legitimate tourist attraction and an object of imaginative marketing.
>
> (Medvedev 1999: 43)

According to Butler (1996), this is probably so because it is something that most people do not experience in their normal lives, for it is contrast from the ordinary that humans seek.

Some border markers are actually designed to be attractive objects of attention and at some crossing points tourists are provided with areas where they can pull aside for a closer view of the boundary. In the presence of a border marker, people have the propensity to stand on it, over it, or in some other way, straddle it. For example, one author confesses his sense of excitement as he approached the US–Canada border: 'the romance of it – this foot in Canada, that one in Alaska – is fetching' (Lopez 1989: 97, quoted in Ryden 1993). According to Ryden (1993: 1), people like to have themselves photographed straddling political boundaries because it is truly the only way they can be in two places at one time. Similarly, a trans-continental drive would probably reveal several motorists stopping at state or provincial borders to photograph 'Welcome to ...' signs to document their vacations (Plate 3.1).

Plate 3.1 This couple photographed themselves at the welcome to Arizona sign to document their cross-country vacation

Perhaps more importantly, however, is where actual border markers become primary tourism objects in a region, bringing economic benefits into local communities. Although there are fewer subnational border attractions than international ones, several do exist. The best known example in North America is the Four Corners Monument where Utah, Colorado, Arizona, and New Mexico meet. This is the only place in the United States where four states meet in one spot, and the exact location is marked with a large granite, bronze, and concrete monument that 'stands on a tiny corner of each state' (Navajo Parks and Recreation Department n.d.: 2) (Plate 3.2). Thousands of visitors stop by the site every year to stand astride the monument, so that they can claim to have been in four places at once (Ryden 1993). In response, the Navajo Nation's Parks and Recreation Department, the agency which administers the site, has established a visitors center, and Navajo vendors sell handmade jewelry, crafts, and traditional foods. The site is an important tourism resource for all four states and holds a prominent place in each state's promotional literature.

The Maryland–Delaware and North Dakota–South Dakota boundaries in the United States provide other examples of domestic borders that function as tourist attractions. On Fenwick Island (Maryland–Delaware), a border stone dating from 1751 is an important highlight of a visit to this popular east coast resort, and inland other markers are promoted as important heritage attractions in both states' tourism literature (Delaware Tourism Office 1987;

Plate 3.2 This tourist is officially in four places at once – Utah, Arizona, New Mexico, and Colorado

Maryland Office of Tourism Development 1989). What is particularly interesting about these markers is that they marked part of the border (Mason–Dixon Line) between the Confederate South and the Yankee North during the American Civil War (Griswold 1939). Similarly, the original stones from 1891 that mark the North–South Dakota border are promoted as tourist objects by both states, but particularly by North Dakota (North Dakota Parks and Tourism 1992). Owing to their isolated locations and lack of connection to other attractions, the stones likely do not generate large sums of tourist dollars into the Dakotas' economies. Nevertheless, they are curious landmarks in a fairly expansive and mundane environment, and have become substantial heritage monuments worthy of conservation (Iseminger 1991).

County, township, and municipal boundaries are less important tourist objects unless they display unusual welcome signs or names such as Kalamazoo, Michigan (USA); Hell, Norway; or Llanfairpwllgwyngyllgogerychwyrndrobwllllantysiliogogogoch, Wales (Timothy 1995b). Then they become interesting spectacles. Many communities erect border signs that boast of people or events of local importance. For example, at the border of Bowling Green, Ohio (USA), it is made clear to visitors that the community is known for the International Tractor Pulling Festival each summer and that it is the home town of figure skating celebrity Scott Hamilton. According to Zelinsky (1988), the primary role of community border signs is to persuade travelers to pause for a few hours or days, to sightsee, to shop, or to get involved in the community in some other way. Ryden (1993: 2) suggests that the primary purpose of welcoming signs is to 'shape and manipulate the feelings of difference and distinction that we experience when we cross any geographical border' and that states and communities erect signs to make passersby aware of the distinctions between the area they just left and the one they just entered.

International frontier monuments are important icons in some locations, and if peaceful relations exist between neighboring countries, they can become a major focus of a community's tourism development plans. An interesting monument marks the point where Norway, Finland, and Sweden come together. The site is accessible only by hiking trails, and within Finland's Mallan Nature Reserve, it is the most popular attraction (Lloyd and Swift 1993; Suomen Matkailuliitto 1993). In the same way, the border gate between Macau and China is listed in Macau literature as one of the city's primary sites (Macau Government Tourist Office 1994) and is now preserved as an historic monument (Plate 3.3). On the eastern edge of the city of Shenzen, China, the large town at the Hong Kong border, lies the small community of Shatoujiao. In the village is a famous street named Sino-British Street, which is bisected by a row of boundary stones, on the sides of which are engraved the names of the two governments in power at the end of the last century. Prior to the re-unification of Hong Kong and China in 1997, this town had become a must for Chinese visitors to Shenzen because it was

Plate 3.3 Until the mid-1990s, this gate was the pedestrian entrance into China from Macau. When this photo was taken in 1994, it had recently been sealed with a sheet of thick plastic, pedestrian traffic was re-routed around the gate, and the gate had achieved heritage monument status

'perhaps the only place where one [could] literally straddle the border between Hong Kong and China' (Cao and Ge 1993: 9).

On the US–Canada border a concrete peace arch was constructed in 1920 to commemorate the good relations between the two countries. Today, on the US side, the monument and its surrounding area have been designated a state park in Washington. Interpretive trails, picnic areas, shops, and an annual Peace Arch Day festival all form the basis of Blaine, Washington's tourism economy.

Even borders between hostile neighbors can be tourist attractions. Although the demilitarized zone between North and South Korea is still a point of contention between the two countries, it has become a major tourist destination. Every week, busloads of foreign and South Korean tourists stream from Seoul into Panmunjom, the village bisected by the border where the treaty of armistice that halted the Korean War was signed (Pollack 1996a; 1996b). Panmunjom is also a popular destination for tourists in North Korea (Hall 1990b).

Former boundaries that no longer function in their original capacity, but which are still visible in the cultural landscape, are known as relict boundaries (Hartshorne 1936). Examples of these abound in nearly all parts of the world. Perhaps the most widely recognized ancient example is the Great Wall of China, which was built between 246 and 209 BC as a fortification between China and Mongolia. The wall is one of China's leading tourist attractions, and a trip to Beijing would hardly be considered complete without at least a one-day tour to view and hike along this magnificent structure (Toops 1992). Hadrian's Wall is another good example of an ancient relict boundary that is now a major tourist site (Plate 3.4). It was started in AD 122 and completed within a few years by Roman Emperor Hadrian as a method of marking the northernmost frontier of the Roman Empire. The wall is considered to be the most impressive and significant remnant of ancient Rome in Great Britain. It was designated a UNESCO World Heritage Site in 1987, and a great deal of work has recently been undertaken to manage, map, delimit, and conserve the site (English Heritage 1996; Turley 1998).

A more recent example is the Berlin Wall and the entire East–West Germany divide. During the period when this frontier functioned as a major barrier to travel and tourism, it also functioned as a leading attraction, where elevated observation platforms on the western side allowed people to view fortifications and communities in the East (Koenig 1981; Maier and Weber 1979). With the demise of the East–West divide in 1989–90, it became even more of a tourist attraction. In fact, this relict border underlies much of the tourist appeal of today's Berlin. Checkpoint Charlie, a West Berlin border station and perhaps one of the best known artifacts of the cold war, is now featured in the new Checkpoint Charlie Museum, which is visited by more

Plate 3.4 Hadrian's Wall is one of the best remaining Roman sites in Britain. It is an excellent example of a relict boundary that attracts significant tourist attention

Source: Reprinted with the permission of Dr Nathan Richardson

than half a million tourists each year and which also houses other cold war artifacts from the old frontier zone, including parts of the wall itself (Borneman 1998; Kinzer 1994; Light 2000), and pieces of the wall are still being sold to tourists. Scars of the old border in Berlin and throughout the countryside are disappearing rapidly. Intense building construction is filling the empty frontier zone in Berlin, but in the countryside, much of the old border still lies as wasteland or is being used as areas of natural vegetation regeneration. As a result, some parts of the border in the countryside, including guard towers and wall fragments, have now assumed heritage status and form part of a *Grenzland Museum* (Borderland Museum) (Blacksell 1998). In town, several segments of the wall are being preserved and have become key sites on tours of the city (Finn 2000; Light 2000). Owing to rapid urban development in Berlin, it is becoming more difficult to find the exact location where the wall used to stand. This normally would not bother most local residents, but tourism centered around the old border has become so important economically for the city that 'Berlin helps make ends meet by luring tourists who want to catch a last glimpse of the cold war and whose first wish is to see where the wall was. So, to satisfy them, the city has devised a new east–west border – a red stripe painted through Berlin's heart along the route of the demolished wall' (*Economist* 1997: 56).

Border theme attractions

In addition to the actual borders that function as tourist objects, several examples exist of attractions that focus on the theme of international boundaries. The Borderlands Museum in Germany is a good example of this. Although several types may exist, parks and trails are the most common border theme attractions. The International Peace Garden (IPG) is one of the best examples in North America. In 1932, this botanical garden was formed straddling the border between North Dakota (USA) and Manitoba (Canada). It was strategically placed in the center of the continent halfway between the Atlantic and Pacific coasts and only 49 kilometers north of the geographical center of North America (International Peace Garden n.d.). North Dakota and Manitoba both donated adjacent tracts of land totaling 930 hectares for the purpose of establishing the park whose primary purpose is to commemorate the peaceful relationship between the two countries and to stand as a monument to worldwide peace. Each summer more than 150,000 flowers are planted and ceremonies are held that commemorate peaceful relations across the border (Mayes 1992). The IPG is now one of the most important tourist attractions in both Manitoba and North Dakota and is featured prominently in the tourism literature of both regions. Approximately a quarter of a million people visit the Garden every year from all parts of North America and overseas (Timothy 1999a).

A recently established nature hiking trail criss-crosses the borders of Norway, Sweden, and Finland over a dozen times for a distance of 800 kilometers. One Finnish brochure calls the experience an international 'hike without borders' through the far north (Suomen Matkailuliitto 1993). Borders and borderlessness are a major focus of this summertime destination. Along the US–Mexico frontier an international heritage trail, *Los Caminos del Rio*, runs for over 300 kilometers from the twin cities of Laredo and Nuevo Laredo to the Gulf of Mexico. The trail, comprised of US and Mexican highways weaving back and forth across the frontier, links dozens of historic sites of common interest to both sides, including Spanish colonial settlements, historic ranches, border towns, natural landscapes, battlegrounds of border raids, Mexico's War of Independence, and the United States Civil War (Jones 1997; Sánchez 1994; Steffens 1994).

Non-political boundary lines

Lines of time and position on the earth, while not necessarily forms of political boundaries, are borders between temporal and spatial elements of the human experience. Demarcated lines of longitude and latitude signify temporal differences between, and define the locations of, places. Like borders, these lines in many cases wield significant tourist appeal in communities throughout the world.

The Equator, for example, is an important attraction for tourists. A road sign on interstate highway 91 just south of the US–Canada border in Derby Line, Vermont, announces boastfully that travelers have just crossed the 45th degree of north latitude – the halfway point between the Equator and the North Pole (Timothy 1998a). In Kenya and Uganda, large signs stand where the Equator intersects the main north–south highways (Love 1998). At the sites guides explain the cosmic significance of the zero degree line of latitude to tourists. Likewise, a popular guide book on Southeast Asia suggests the following action to potential tourists on the Indonesian island of Sumatra: 'A huge globe by the roadside north of Bukittinggi marks the equator. Spend 5 minutes hopping back and forth across the line and you'll be able to say, "the equator? Oh I've crossed it dozens of times"' (Turner 1994: 244).

The Arctic Circle is another important tourist attraction in most of the countries it bisects, at least at points where it is marked on or near highways for tourism purposes (Plate 3.5). In one Alaska state holiday guide the intrigue affiliated with crossing the Arctic Circle is promoted to potential visitors as follows: 'For some visitors, the thrill of crossing the Arctic Circle is the most memorable moment in their Alaska journey' (Alaska Department of Commerce and Economic Development 1989: 92). Similarly, promotional material from Yukon Territory, Canada, romantically reminds travelers that 'From this line north the sun never sets on the summer solstice ... and never rises on the winter solstice' (Western Arctic Tourism Association, n.d.:

Plate 3.5 'Hanging out' at the Arctic Circle is a popular activity for tourists visiting this location just north of Rovaniemi, Finland

2). Just north of Rovaniemi, Finland, the Arctic Circle has been the focus of much promotional effort. The circle is marked by a sign in several languages, and a Santa Claus village shopping and crafts center has been built. The spot is an important attraction, and tourists can even buy certificates, which prove that a person has crossed the Arctic Circle (Pretes 1995; Timothy 1998a). An underground theme park near the Arctic Circle, known as Santapark, opened for business in November 1998. In addition to the shopping and crafts center already available at the Circle, the new park boasts rides, shows, and a workshop with elves (*USA Today* 1998).

Just as interesting are lines of longitude. Every year, thousands of visitors to the Royal Observatory in Greenwich, England, have themselves photographed standing astride a strip of brass embedded in the ground which marks the prime meridian (Plate 3.6). Every position on the earth is defined by its distance east or west of Greenwich (longitude) and its distance north or south of the Equator (latitude). The prime meridian attracts nearly half a million visitors a year who are 'proudly photographed with one foot in both hemispheres' (*Geographical Magazine* 1992: 5). Local shops and coin-operated vending machines near the line sell certificates issued by the Royal Observatory which state, 'This is to certify that (name of visitor) has today stood simultaneously in both Eastern and Western halves of the World in Greenwich, London.' Owing to this interest by tourists, many of Greenwich's recent tourism development efforts have focused on the prime meridian (Greenwich Waterfront Development Partnership 1995; 1996).

Attractions bisected by borders

Curiosity seekers are sometimes attracted to populated areas where political lines divide cities, villages, and even individual buildings. Texarkana, a medium-sized city in the United States is divided by the border of Texas and Arkansas, hence its name. The border runs down the middle of State Line Avenue and at one point intersects the post office. According to one brochure, 'Texarkana is the home of one of the most unique post offices in America ... With half the building in Texas and half in Arkansas, the post

Plate 3.6 The Prime Meridian in Greenwich, England, can be straddled with one leg in the eastern hemisphere and one in the west

office's official address is listed as Texarkana, U.S.A. 75501' (Texarkana Chamber of Commerce n.d.: 2). In front of the building is photographer's island, where tourists are encouraged to photograph their friends and family members with one leg in each state. Much of the city's promotional efforts center around its borderline location. The city of Lloydminster, Alberta/Saskatchewan, in Canada, boasts a similar situation which is also the focus of the town's promotional efforts (Lloydminster Tourism and Convention Authority n.d.).

The Haskell Library and Opera House straddles the US–Canada border in the divided community of Rock Island, Quebec, and Derby Line, Vermont. In 1995, 924 visitors signed the guest registration book at the library and nearly as many in 1996. According to the librarian, however, more than a third of the visitors do not sign the register, so the actual number of out-of-town visitors is well over 1,000. Although not a large number, it is significant for a community of 2,000 residents. Visitors come from all over the world, usually in conjunction with day trips from Montreal and other nearby areas of Quebec (Ministère du Tourisme 1993). Despite the building's architectural uniqueness and famous opera stage, according to the librarian, the black line painted down the middle of the floor marking the international boundary is the highlight of most people's visit. In fact, people usually spend nearly as much time jumping back and forth across, straddling, and photographing, the border line, as they do enjoying the architectural design and artistic importance of the building. Several other alluring curiosities exist along the US–Canada boundary where taverns, grocery stores, and even private homes are divided (Sanguin 1974; Sobol 1992; Vesilind 1990).

In addition to urban areas, boundaries also divide cohesive regions of cultural significance (Anderson 1994; Madden 1995) and areas of rich natural landscapes (Lucas 1964; Marić 1988; Rennicke 1995; Vorhes 1990) that are extremely important tourist attractions in their respective regions. Rivers are commonly used as borders between nations, and as a result, several waterfalls are bisected by international boundaries. For example, Victoria Falls, which lies on the Zambia–Zimbabwe border, is one of the most important attractions in southern Africa (McNeil 1997). Divided by the borders of Argentina and Brazil, Iguazu Falls is of primary importance as an attraction in both countries, and indeed in all of South America (Lampmann 1997; Salomon 1992). Likewise, Niagara Falls, one of North America's premier destinations, has been attracting tourists for over 200 years (Niagara Parks Commission 1992; Schwartz 1997) and continues to succeed in luring more than 12 million visitors annually to the Canadian side of the border alone (Timothy 1995b). Although the Falls themselves are the primary attractions in these three locations, the boundaries that run through them may indeed add a degree of intrigue to them. To view the Falls from both countries, especially if the points offer different vistas, may be a realistic ambition for many tourists (Nin 1998; Timothy 1995b).

The management of cross-border attractions is difficult at best since they lie in two or more jurisdictions, and thus different management approaches for conservation and promotion are often utilized (Timothy 1999a). Managers on one side of a border may possess an entirely different value system and view resource conservation from very different eyes. The management of Niagara Falls, for example, differs significantly between the US and Canadian sides (Getz 1992; 1993). This issue will be discussed further in chapter 4.

Human fascination with borders – collecting places

All time lines and political divides are fabrications of human experiences and reflections of socio-political values. Maybe this is why so many people find them fascinating, especially tourists, as the examples above and several observers have suggested (e.g. Matznetter 1979; Medvedev 1999; Ryden 1993; Timothy 1995b; 1998a). Lines of time, position on the earth, and political ideology are invisible until they are marked on the ground with tangible objects. Once this is done, they have potential to become tourist attractions.

One motivation for straddling or crossing temporal and political borders, collecting places, was introduced and examined by Timothy (1998a). Much of this discussion is taken from his work. According to this notion, places are not necessarily consumed to the extent that Urry (1995) examined in his work, *Consuming Places*. Rather, collecting places suggests that destinations can be collected in a less consumptive sense. Collecting places refers to a process whereby locations visited are enumerated, and wherein there is a desire to visit additional places for competitive reasons. This competition is usually against other travelers, friends and relatives, and even the travelers themselves. To collect places lends credibility to people's knowledge of the world and provides instant recognition that they belong to some unique group of privileged travelers. By collecting places, people can partially satisfy their basic need for self esteem (Maslow 1954) by seeking the admiration and recognition of neighbors, friends, and relatives.

Collecting places can be viewed in three forms. First, some people cross borders just so they can claim that they have been in a different country or particular place. Wolf's (1979) work on the Germany–Austria border supports this claim. He found that many tourists cross international boundaries just to be able to say, for prestige reasons, that they have been in another country. Second, some people travel to many places so that they can boast of numbers and widespread adventures. This commonly results in phrases such as 'been there, done that' and 'Oh, I've been to (number) countries in Europe.' One magazine article plays into this notion by encouraging tourists to 'stay in Holland's medieval Maastricht and see four countries in one unhurried day' (Keown 1991: 113). Finally, some people collect places solely for the purpose of impressing others by their choice of destinations. For many people, it is not necessarily the number of places they have been, but rather the uniqueness or peculiarity of the destinations they choose to visit. According to Butler (1996: 216), such destinations 'represent attractions to be acquired almost regardless of location, difficulty or cost of access, and often even regardless of risk or hazard. Indeed, the more remote the location, the more valued it is as a collector's item.' Such places might include isolated destinations on the global periphery, such as Greenland or Antarctica, or perhaps places that are feared or forbidden, such as Cuba and North Korea for Americans. This theory might explain why so many western tourists desire to travel to places that have traditionally been off limits. The interest in experiencing the peculiarities and closed nature of North Korea, for example, is a unique push factor for Europeans and North Americans (Foster-Carter 1996). Viken's (1995: 79) discussion on tourism in Svalbard indicates that there are three different tourist types in isolated, arctic regions: the naturalist, the scientist, and the conqueror. Conquerors are those for whom it is of major significance to have visited Svalbard, one of the world's most remote and peripheral regions.

> The conquest may involve different motives and forms ... For some people it is primarily a symbolic act. To come to Svalbard is to consolidate and manifest the fact of being one of the masters of the world. For others it is the status which is most important. A trip to Svalbard is a sign of being a great traveller.
>
> (Viken 1995: 79)

Destinations that possess socio-political and economic systems that are different from that of one's own society appear to be common targets for place collectors. For example, thousands of Americans circumvent US government restrictions on travel to Cuba every year by traveling to the island via Mexico, the Bahamas, or Jamaica, just so they can say that they have visited Cuba (Reynolds 1997). By the same token, Viken *et al.* (1995: 104) concluded that, prior to the collapse of the Soviet Union, the main reason for traveling to the USSR as a tourist from the West was the very fact of going there.

Together these concepts constitute collecting places. Such a theory might help explain people's desire to visit sites of geodetic linear importance and straddle them. To possess the experience of having been in two places at once, having stepped into a different country or hemisphere, having added another country or location to one's list of places visited, or of having set foot into a place that is unique and unusual, is to be recognized and admired. Travelers who seek to fill their passports with frontier stamps and visas, photograph themselves straddling geodetic and political lines, or 'Welcome to ...' signs on major highways, do so in an attempt to document the places they have collected (Timothy 1998a).

Borderlands

Border regions are areas in the immediate vicinity of international boundaries whose economic and social life is directly affected by proximity to a border (Hansen 1981). As mentioned earlier in this chapter, borders create environments conducive to the development of certain types of tourism because of political, cultural, and value differences on opposite sides (Matley 1976; 1977). When frontiers mark the interface between different politico-cultural systems, prices of goods, and policies and regulations pertaining to certain licentious diversions, activities tend to develop in response to foreign demand. The following sections discuss several tourist activities that have developed in borderlands owing to their proximity to political frontiers.

Cross-border shopping

One of the most ubiquitous forms of borderlands tourism is cross-border shopping, traditionally known in marketing literature as outshopping. This activity, wherein people cross political boundaries to purchase goods and services in foreign jurisdictions, is common all over the world. Several scholars have examined this phenomenon in Europe (e.g. Bachvarov 1997; Bygvrå 1990; Fitzgerald *et al.* 1988; Hajdú 1994; Hall 1991a; Joyce 1990; Kovács 1989; Leimgruber 1988; Minghi 1994b; 1999; Sándor 1990; Weigand 1990), the Americas (Asgary *et al.* 1997; Chatterjee 1991; Di Matteo and Di Matteo 1996; Mikus 1994; Patrick and Renforth 1996; Timothy and Butler 1995), and Southeast Asia (*Economic and Business Review Indonesia* 1993; Ngamsom 1998), although relatively few of these discuss the subject in the context of tourism.

While many characteristics of cross-border shoppers resemble traditional tourists, the primary difference in behavior is that the former are motivated to purchase general merchandise from establishments like department, grocery, outlet, and variety stores (Richard 1995; Weigand 1990). However, it should be pointed out that many of these establishments are also important amenities in traditional tourist destinations, and many cross-border shoppers also purchase things of a more touristic nature such as souvenirs and books.

Leimgruber (1988: 54) argued that four conditions must be present for border shopping to take hold and develop. First, there must be sufficient contrast between the home environment and that on the 'other side.' This is usually viewed in terms of differences in price, product quality, and selection. Second, residents of one country have to be aware of what is on the other side. Populations must have adequate information on the goods offered beyond the customs barrier either through the media (advertising) or through personal visits. Third, potential shoppers need to be able and willing to make the trip, especially taking into account such things as exchange rates and personal mobility. Finally, the boundary must be sufficiently permeable. At the risk of stating the obvious, it is clear that borders with fewer restrictions will become the locations of choice for outshopping rather than those with extensive formalities and barricades.

More specifically, several factors contribute to the growth and development of cross-border shopping (see Table 3.1). First, exchange rates between currencies appear to be the most significant reason for the rise of this phenomenon (Richard 1995). Research has shown causal relationships between exchange rates and levels of international travel generally, as well as international outshopping in particular (e.g. Di Matteo 1993; 1999; Di Matteo and Di Matteo 1993; 1996; Diehl 1983; Gibbons and Fish 1987; Timothy 1999b). When exchange rates are favorable to Country A, residents will travel to Country B to shop. For example, Figures 3.1 and 3.2 illustrate in graphic form the results of a simple regression analysis (Timothy 1999b) aimed at testing the relationship between cross-border shopping and exchange rates in the United States and Canada. These illustrations demonstrate a strong inverse relationship between levels of cross-border shopping and exchange rates, and Figure 3.2 particularly shows on a monthly basis that even the slightest shifts in the value of the Canadian dollar to the US dollar resulted in significant and rapid changes in numbers of Canadian shoppers during the period 1985–97.

Second, like exchange rates, when taxes in one area are higher than in a neighboring region, people from the high-tax area will travel across the border for cheaper goods and services. Low sales, liquor, tobacco, and gasoline taxes all contribute to outshopping to areas where these are lower or nonexistent (Bygvrå 1990; Hidalgo 1993). Third, distribution channels in smaller countries are sometimes not as efficient or competitive as in larger states, with

Table 3.1 Factors contributing to the growth of cross-border shopping

- Favorable exchange rates between currencies are a major motivation for shopping in countries where currencies are weak.
- Higher taxes on one side of a boundary drive people across where tax rates are lower.
- As a result of economies of scale, small distribution channels, and a lack of competition in smaller economies, higher profit margins exist, driving up the cost of consumer goods at home.
- There is often a wider selection of goods and services across the frontier.
- Customer service in neighboring jurisdictions may be better than at home.
- Many people shop abroad owing to differences in the opening hours and days of shops, particularly on weekends and holidays.
- Shopping abroad is entertaining and enjoyable. For many people, the thrill of crossing a border is compounded by the assortment and quality of products available.

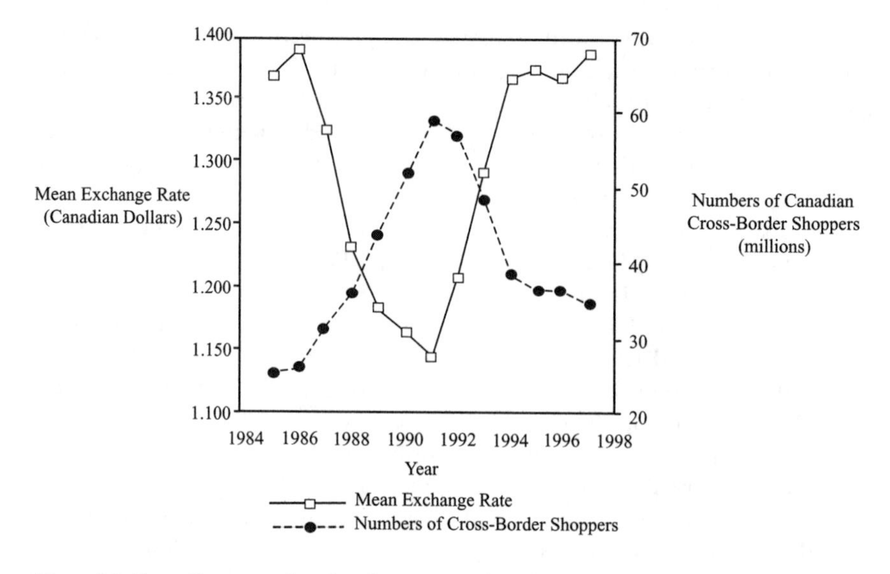

Figure 3.1 Canadian cross-border shoppers and exchange rates
Source: Statistics Canada 1986–98; Bank of Canada 1985–98

the result that profit margins are usually higher in order to make up for the limited market base (Canadian Chamber of Commerce 1992). A fourth reason is a wider selection of goods, particularly in terms of fashions, brand names, and product quality. In situations where one country has a smaller population, or market, product lines have traditionally been less varied than

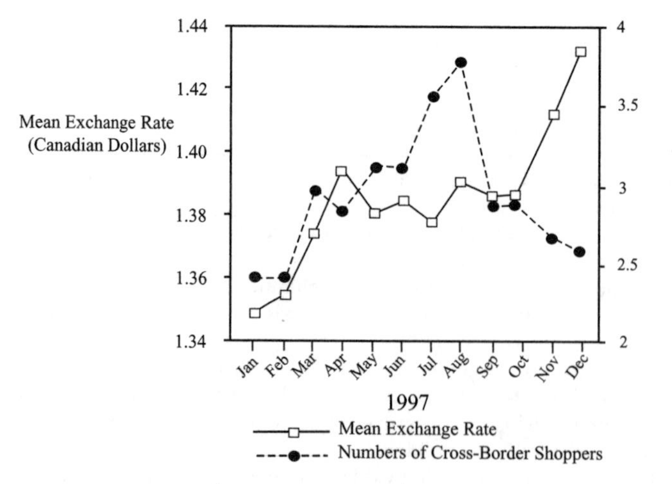

Figure 3.2 Canadian cross-border shopping by month, 1997
Source: Statistics Canada 1986–98; Bank of Canada 1985–98

those in adjacent nations with larger market bases (Government of Ontario 1991). Customer service is the fifth factor. Many people report enjoying the level of service and individual attention shown them in neighboring countries (Canadian Chamber of Commerce 1992; Government of New Brunswick 1992). The sixth factor, different opening hours, also attracts foreign consumers, such as Sunday shopping in states or countries near other areas where this is not permitted (Timothy and Butler 1995). Finally, that a shopping trip is pleasurable and can function as a family vacation or other recreational activity is also an important motive (*London Free Press* 1992).

Although much of this activity is motivated by economics, it is often viewed as pleasurable and can sometimes act as a sort of a mini vacation (Goodman 1992; Timothy and Butler 1995). A study by the Canadian Chamber of Commerce (1992) shows that one of the main reasons Canadians shop in the US is because it allows them to get away from home and experience a change. Furthermore, 'their trip is a family outing and, depending on the length of their stay, may even be seen as a vacation' (Canadian Chamber of Commerce 1992: 6). As with consumer activities in many other contexts, border shopping has taken on a pleasure, or recreational, element. It is no longer viewed as merely an economic activity that must be carried out as part of a mundane set of daily chores. According to Timothy and Butler (1995), the fact that cross-border shopping experiences seasonality, is an indicator that many people participate in the same manner as in other forms of recreation. A lack of seasonal variation might imply a more purely economic motivation. The fact that many border shoppers eat out in restaurants and visit museums, movie theaters, historic sites, and sporting events while abroad, indicates that the trip does in fact include other forms of pleasure seeking (see Plate 3.7). By extension, although border shoppers are day visitors or excursionists, the activity itself may be considered a form of tourism because people cross a border, spend money, are often motivated by enjoyment, and use other services. Some people even spend the night so that they can claim a higher tax-free allowance. Several authors have recognized the role of cross-border shopping in tourism (e.g. Johnston *et al.* 1991; Murphy 1985; Ryan 1991). In some areas, it has become so popular as a recreational and tourist activity that guides have been published to show potential shoppers the best locations for the best bargains (e.g. Cahill 1987; Meldman 1995; Szabo 1996; Yenckel 1995).

Two additional types of international consumerism exist that are not included in the general model of cross-border shopping: that done by visitors already on vacation and mail-order catalogue shopping. Vacationers tend to buy small items that can be transported home with relative ease, but they may also buy high-value goods since their duty-free allowances nearly always increase with time spent abroad. Some mail-order customers may never visit the country where goods are purchased, but shop by mail, telephone, and the Internet for the same reasons that motivate other cross-

Plate 3.7 This sign, located about ten kilometers inside Canada at Woodstock, New Brunswick, attempts to entice Canadians to shop and dine in Houlton, Maine, USA

border shoppers – low prices, wider selection, and a leisure experience (Timothy and Butler 1995: 29).

Cross-border shopping demonstrates certain spatial characteristics. For example, Jansen-Verbeke (1990: 130) concluded that the further a person travels for shopping, the more likely it is that the trip has a leisure-based nature. Similar conclusions were drawn by Ritchie (1993) and Timothy and Butler (1995). These studies also show how the frequency of shopping and types of goods purchased are influenced by how far people travel to shop across a border. According to Timothy and Butler's (1995) model (Figure 3.3), the further people live from the frontier the less frequently they will cross, but the value of goods purchased on each trip will likely be higher. From an economic perspective, people who live in the proximal zone, up to 50 kilometers from the border, are more inclined to cross frequently and are willing to go for everyday items such as gasoline, beer, tobacco, groceries, and restaurant meals. Consumers who live in the medial shopping zone (50–200 kilometers) cross the border less often and tend to buy higher-value goods, such as clothing and tools. Residents of the distal zone (beyond 200 kilometers) cross least often and are inclined to purchase big-ticket items, such as furniture and appliances. Similarly, Mikus (1994: 444) concluded that the higher the price differential across the border, the further consumers are willing to travel – a concept also discussed by Fitzgerald *et al.* (1988). This

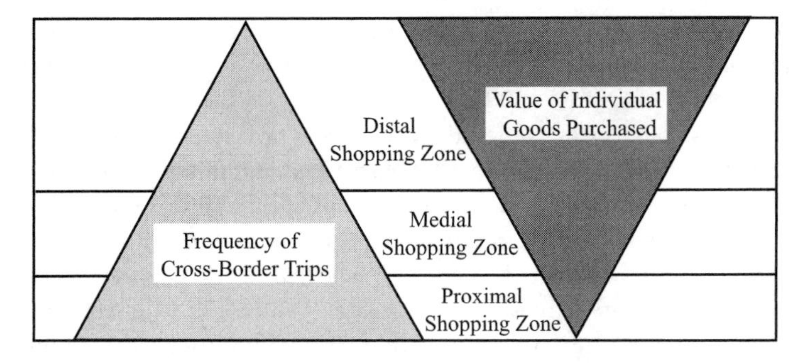

Figure 3.3 Spatial characteristics of cross-border shopping
Source: after Timothy and Butler 1985. Reprinted with permission of Elsevier Science

may be the general trend, but it must be recognized that obvious variations of this pattern exist. Proximal residents are probably just as likely to purchase big-ticket items on one of their frequent trips, and when distal- or medial-zone residents shop they are probably just as likely to eat out and buy gas and groceries while abroad as residents living closest to the border.

Many of these concepts can be demonstrated empirically by examining cross-border shopping along the US–Canada and US–Mexico borders. Tourism has always been one of the most important areas of international trade between the United States and Canada, and shopping has long been one of the most important activities undertaken by visitors, particularly by Canadians in the USA (Dilley *et al.* 1991; Kreck 1985). Canadians traveling to the United States specifically to shop was a fairly consistent phenomenon throughout the early 1980s at around 20–25 million trips per year. However, in 1987 the phenomenon took off at an accelerated rate reaching its peak in 1991 (see Figure 3.1), owing largely to the appreciation of the Canadian dollar against the US dollar, which began in 1986 (Di Matteo 1993). As Figure 3.1 shows, almost 31 million Canadian shopping trips were made, but that number nearly doubled by 1991 when over 59 million cross-border shopping trips were taken. In 1992, the number of trips began a consistent decline until in 1997, when just under 35 million trips were made. In an absence of precise data, same-day automobile trips and expenditures are commonly used by government agencies as representative measures of cross-border shopping (Di Matteo 1993; Richard 1995). Thus, the same definition is adopted here.

In its heyday this shopping craze was widely blamed for the loss of thousands of jobs and retail bankruptcies in Canada. It has also been blamed for millions of dollars in lost revenue for national and provincial governments. Some estimates placed the number of lost Canadian jobs at around 55,000 and Can.$3.5 billion in lost sales in 1991 (Government of New Brunswick 1992). The financial loss to the province of Ontario alone was estimated to be around Can.$1 billion the same year (Chamberlain 1991; Government of Ontario 1991). Canadian border towns were hit especially hard, although the effects were felt all over the country (Kemp 1992). Congestion and long delays at the border in both directions were also cited as critical negative impacts (Government of Ontario 1991).

While Canadian businesses suffered, the US borderlands flourished (Kendall and Kreck 1992; Kreck 1985). In North Dakota alone, the impact of Canadian sales was estimated at nearly US$2.5 billion in 1991, amounting to 52.4 percent of the state's total retail sales (Goodman 1992), and many individual US businesses along the border reported sales comprised of between 50 and 80 percent Canadian patronage (Scanian 1991a; 1991b). In fact, the activity became so popular that Canadian tour companies began offering one-day shopping trips, which they eventually expanded to include two- and three-day shopping tours as well. These remained popular even after the 1992 descent of the Canadian dollar began (*Kitchener-Waterloo Pennysaver* 1993).

Research has shown that the primary reason Canadians shop south of the border is that prices are generally lower in the United States, caused by lower profit margins, more retail competition, lower taxes, and earlier in the 1990s, the strong Canadian dollar (Chatterjee 1991; Government of New Brunswick 1992; Government of Ontario 1991; Stevenson 1991). A wider variety of goods, better service, active promotion, free amenities, and businesses accepting Canadian currency at par with the US dollar – a saving ranging from 10 to 20 percent depending upon exchange rates – also contributed to the situation (McWhirter 1992). Sunday shopping in the US was another influential factor. One Ontario study found that 74 percent of that province's consumers who shopped across the border on Sundays said they would shop at home if stores were open that day (Timothy and Butler 1995: 24). A special factor for many Canadians has been the lack of enforcement of tax collection at the border by Canada Customs. Although Canadians were not permitted to import any taxable items duty free if their stay outside the country was less than 24 hours, many found that they could routinely make day trips and relatively small purchases without being charged tax on them at ports of entry.

Media coverage also contributed to the rapid growth of this phenomenon by suggesting that all goods were cheaper in the US than in Canada (Lewis 1990). American businesses fuelled this information campaign by advertising intensively in Canadian newspapers and on television and billboards (Dilley and Hartviksen 1993; Fisher 1990; Stevenson 1991), and many hotels, shopping malls, and stores began advertising their acceptance of Canadian currency at par (*London Free Press* 1992). The introduction of the new seven percent Goods and Service Tax (GST), which replaced an invisible manufacturing tax on a limited range of products, by the Canadian federal government in January 1991, has been denounced as the most unpopular sales tax in Canadian history and has been a major contributor to cross-border shopping (Timothy and Butler 1995). While the GST did prove to be an annoyance to most Canadians, the underlying reason for the rapid increase in cross-border shopping was the 'appreciation of the Canadian dollar, combined with the rise in real per capita income during the boom of the late 1980s, brought about the surge in Canadian expenditures in the United States. The GST merely provided an additional boost to the cross-border spending phenomenon' (Di Matteo 1993: 60), a sentiment expressed by Boisvert and Thirsk (1994) as well.

In their examination of Canadian cross-border shopping, Timothy and Butler (1995) suggested that the phenomenon was not purely an economic activity, rather it was for many people based on a pleasure, or leisure, motivation as well. Ritchie's (1993) research confirms this assertion demonstrating that a third of cross-border consumers were motivated by pleasure, although this number is likely higher since the way his questionnaire was written, participants were required to choose between pleasure and shopping as the main purpose of the trip.

In an effort to keep Canadians at home, the national government made a few significant strides to redirect the flow of dollars across the border. First, it launched a promotional campaign in 1992. Playing on sentimentalism and patriotism, the Can. $3 million campaign focused on the beauty and heritage of Canada and attempted to make Canadians 'fall in love with their country' (Hodgkinson 1992: B13). Extensive television and billboard advertising, much of it promoting shopping in Canada, could be seen in many communities as part of the government's stay-home crusade (Timothy and Butler 1995). Likewise, unable to offer much in the way of financial breaks, storekeepers and mall owners attempted to incite patriotic feelings on the part of Canadians in hopes that this alone would cause them to shop in Canada. Malls in London, Ontario, for example, posted signs on doors, which stated 'Thank you for shopping in Canada.' Cross-border shoppers were also commonly accused of being unpatriotic to Canada and contributing to the decline of the national economy (Ahmed and Corrigan 1995); however, it was realized that most consumers were more concerned about their own economic well-being than their national and local economies (Tourism Canada 1992).

Second, bending to heavy pressure by public officials, business people, and residents, legislators in Ontario voted in favor of Sunday shopping, and early in the summer of 1992 it became legal in that province (Timothy and Butler 1995). Finally, the Canadian government initiated some significant changes at the border. For example, Canada Customs began to crack down on the importation of many products by individuals, and tariffs were removed from a range of imported goods enabling merchants to offer lower prices. As a result, Canadian retailers began to narrow the price gap on items that were previously cheaper in the United States (Rinehart 1992). In addition, some provinces, in cooperation with Canada Customs, began to collect provincial sales tax at the border as well.

As noted in Figure 3.1, cross-border shopping began a notable decline in 1992. Although efforts on the part of the government and business community to keep Canadians from leaving might have contributed to this, it was the depreciation of the Canadian dollar, beginning in 1992, that really initiated the decrease in cross-border shopping. The Canadian dollar hit a record low in 1998 (nearly 1.6 to the US dollar) as instability in Asian currency markets and falling prices for wood and other staple Canadian exports drove the country's currency downward (Schneider 1998). This trend has induced record-breaking streams of day trips to Canada by US residents in search of big-ticket items like furs, crystal, televisions, and in some cases, summer cottages (Bondi 1997; Sloan 1998), as well as smaller items like food, diapers, CDs, and clothes (Associated Press 1998). As one US resident put it on his way to a mall in Canada, 'I live here in Detroit, but I haven't been in Canada for years. But hey, the rate is good and I want some new clothes. Better take advantage of it now' (quoted in Bondi 1998: 1). A complete turnaround from 1991, US tour organizers are now bringing busloads of Americans to shopping malls in Canada. According to the author's contacts in London, Ontario, it is common to see busloads of people from Michigan shopping at the city's malls, particularly on weekends.

As early as 1992, this reverse shopping pattern became noteworthy and began benefiting Canadian businesses, whose owners now feel that it is payback time for the drastic losses they endured a few years earlier (*Alberta Business* 1992). Table 3.2 highlights the magnitude of American cross-border shopping in Canada. In 1992 just over 19 million same-day automobile trips were taken to Canada, and in 1997 the number increased to over 25 million. Now American businesses are suffering grave losses. Many stores near the Canadian border have shut their doors permanently, and unemployment levels are high. One store manager interviewed by the author in Houlton, Maine, lamented the difficulties facing US merchants as they deal with the loss of a significant portion of their market base. In an effort to alleviate some of the problem, he sends fliers across the border every week to neighboring Woodstock, New Brunswick, to promote special sales.

Table 3.2 American day trips (cross-border shopping) to Canada, 1985–97

Year	*Number of trips*	*Year*	*Number of trips*
1985	20,522,103	1992	19,019,020
1986	22,555,457	1993	19,012,837
1987	22,278,161	1994	20,667,063
1988	21,465,298	1995	22,745,610
1989	20,795,542	1996	23,803,557
1990	20,691,693	1997	25,252,250
1991	19,808,659		

Source: Statistics Canada 1986–98

The current business climate in American border towns is dismal, and elected officials and business leaders are scrambling to figure out ways to re-increase the flow of Canadian shoppers. Ahmed and Corrigan (1995) suggested several methods to combat this problem in North Dakota. These include refunding state sales tax to Canadian visitors; offering promotional gifts such as coupons for gasoline, meal discounts, and other retail items; increasing convention and visitors bureau membership base; joint marketing efforts with nearby communities; and developing a currency-at-par system, which would decrease profits on a sale-by-sale basis but would increase overall profits through mass consumption.

Canadian tourist literature is beginning to reflect this complete turnaround in shopping flow pattern. The title of one brochure, for example, slyly manipulates the first line of the American national anthem. The brochure, 'Oh, Say Can You Save!' (New Brunswick Tourism 1997), aims to lure US residents to shop in Canada, listing dozens of retail businesses in New Brunswick and emphasizing the favorable exchange rate for Americans.

In addition to shopping, all forms of American travel to Canada have seen a steep increase in the past few years, as the costs of hotels, restaurant food, transportation, and organized tours have decreased as well (Sloan 1998; Stinson and Bourette 1998). According to a restaurant owner in one Canadian border town, business has increased between 30 and 40 percent since mid-November 1997, when the Canadian dollar began its biggest downward spiral. 'We like the Americans because they love to spend. They are the ones ordering $40 to $50 bottles of wine, the lobster tails, the surf and turf – all of the most expensive things on the menu' (quoted in Stinson and Bourette 1998: 1). Interviews by the author with restaurant owners and customers in Boissevain, Manitoba, revealed that many Americans are now traveling 40–50 kilometers on a regular basis just to eat at Canadian restaurants. For Americans, the favorable exchange rate is extending their action space onto the Canadian side of the border. As one woman put it, 'We are sort of isolated where we live in North Dakota. With the exchange rate we can now afford to come up here for dinner, and the Chinese food is even better.'

Similar conditions exist along the US–Mexico border. Mexican border towns are very popular among American tourists, and most of their purchases include souvenirs and related commodities, such as leather, blankets, sombreros, paintings, piñatas, and colorfully painted trinkets. Most US border zone residents, however, shop for entertainment, physician and dental services, auto repairs, haircuts, tropical fruit, baked goods, liquor, soft drinks, furniture, and prescription drugs (Appleby 1995; Fernandez 1977). Perhaps the most important aspect of cross-border shopping there, however, is Mexicans shopping in the United States. Brown (1997: 115) estimated that border purchases in 1995 by Mexican shoppers in the USA were about US$20–22 billion and accounted for more than one million American jobs. The most common products purchased by Mexican consumers are food and beverages, clothing, footwear, beauty and personal care products, photo and sporting supplies, records, magazines, machinery and tools, and household goods (Pavlakovic and Kim 1990: 9).

In 1976 Mexico devalued its currency by 50 percent in an attempt to stimulate exports and impede imports, and it has done the same thing several times since – in 1982 by nearly 200 percent and again in December 1994 (Patrick and Renforth 1996; Prock 1983). The effects of these moves on US border communities has been the same as those that occurred in the US communities along the Canadian border when that country's dollar began to lose value. At a time when roughly 90–95 percent of all commerce in many US border communities was Mexican (Austin 1979), the 1982 devaluation crippled South Texas merchants by slicing retail sales as much as 80 to 90 percent (Diehl 1983: 122). This event was followed by dozens of businesses closing their doors and unemployment rates rising to nearly 25 percent in some Texas communities. Similar negative effects resulted from the 1994 peso devaluation, since which time the peso has continued to slide. Business is picking up again, and even poorer Mexican consumers are continuing to purchase household staples in the United States (Wasserman 1996). Nevertheless, consumer activity in most US border towns has not reached its pre-1994 level. As a result, towns such as El Paso, Texas, are suffering severe setbacks, and many businesses are declaring bankruptcy (Appleby 1995; Wasserman 1996).

According to one study (Asgary *et al.* 1997), 49 percent of the Mexican shoppers in Texan cities were motivated primarily by cheaper prices. The remaining 51 percent found other reasons, such as quality, variety, availability, and service, to be particularly appealing. In common with Di Matteo and Di Matteo's (1996) findings about Canadians, when per capita income increases so does Mexican spending in the United States (Asgary *et al.* 1997). Not all Mexican shoppers are motivated completely by economics or availability of goods, however, for as the Asgary *et al.* study shows, sightseeing, eating in restaurants, and visiting friends and relatives were also important reasons for crossing the border.

The underlying cause of such high levels of Mexican consumerism in the United States is the existence of *maquiladoras* (foreign-owned factories and assembly plants) on the Mexican side of the border. In 1965 Mexico's border industrialization program was initiated as a means of meeting the employment needs of borderland residents (Dillman 1970a; Holden 1984). As a result, hundreds of these plants have been built all along the 3,400-kilometer-long frontier, and hundreds of thousands of people have moved from the Mexican interior to the northern border in search of better-paying jobs on *maquiladora* assembly lines (Brown 1997; Dillman 1976; Stoddard 1987). Mexicans who prove that they live in border towns can get passes allowing them to cross into the US for stays up to 72 hours each time (Smith and Malkin 1997: 66) – a privilege that supports cross-border shopping.

A unique form of cross-border shopping, barter tourism, appeared along the Russia–China frontier in the late 1980s between the cities of Heihe (China) and Blagoveshchensk (USSR, now Russia), both located on opposite banks of the Amur River. With the normalization of relations between the Soviet Union and China after years of hostilities, their common boundary was opened up to trade in 1988 – an action that permitted the growth of cross-border tourism (Yu 1992). In September of that year, the two sides initiated an exchange of tourists for one-day excursions – a move fully embraced by residents because of their desire to see the outside world (Vardomsky 1992; Zhao 1994a). To avoid the problem of foreign currency exchange, which was still rigidly controlled by both governments, arrangements were made so that each side received the same number of tourists as its counterpart, usually along the lines of 200 people a day, and provided the same services of comparable quality, including food, accommodation, and sightseeing (Zhao 1994b). Each party charged the tourists from its own country for the cost of hosting groups from across the border, thereby eliminating the need for travelers to deal with foreign currencies (Zhenge 1993). In 1988 each side sent 520 people across the border. In 1989 this increased to 8,000 and in 1992 the number rose to 49,000 (Zhao 1994a: 66). At first, tourists from both countries only did sightseeing on these day trips, involving a small amount of shopping using goods from their own countries as the medium of exchange. However, with the collapse of the USSR, civilian trade between Russia and China developed rapidly, and cheaper prices south of the river began attracting Russian shoppers, thus giving rise to the present form of barter tourism (Nin 1994; Zhenge 1993). As the phenomenon developed, it centered less on sightseeing than on the exchange of goods, although sightseeing is still important because the curiosity factor still exists. Chinese travelers carried their goods across to trade for Russian goods, and vice versa. As news of the two cities' success spread, several other twin border towns in the region began to institute similar exchanges, and tourists from the interior also became involved (Zhao 1994a). People from the interior board the train in Harbin in the evening and arrive in Heihe early the next morning. On the Chinese side of the river they purchase lighters, cosmetics, jackets, clothing, and other goods that are popular among Russians. These things are then exchanged for Russian goods, such as furs, watches and leather products, when the Chinese go sightseeing in Blagoveshchensk the next day (Zhenge 1993: 71).

Barter tourism is still important in the region (Gengxin 1997; Gormsen 1995; Kotkin and Wolff 1995; Roehl 1995), and it has changed the image of Heilongjiang Province from that of an underdeveloped and peripheral place to one of growing prosperity and international trade. From January to April 1997 the import and export volume of border trade in China's Heilongjiang Province alone totalled US$220 million, an increase of 133 percent over the same period the year before, and efforts are under way by local governments to strengthen and quicken the development of border trade (Gengxin 1997). However, frontier residents suffered a setback when, in 1994, Moscow canceled visa privileges and tightened some restrictions to the extent that border trade in the Amur Oblast, which amounted to US$416 million in 1992, decreased by a factor of four in 1994, and in 1995 it fell by another third (Belyy 1996).

It is important to note that this activity developed at the time when hard currency was difficult to acquire, so that the reciprocal exchange program instituted initially made it possible. However, the restructuring of the Russian economy into a free market means that obtaining foreign currency is no longer a challenge. Russians are willing to pay US dollars to travel to China, something that continues to create a challenge for their neighbors to the south who are still restricted by currency regulations (Zhao 1994b: 403).

As a result of this booming trans-boundary trade, several prominent people in the region are lobbying in China for special status and national benefits to their cities to match the achievements of Shenzen and other cities designated as special economic zones (Rozman 1995: 281). Furthermore, border trade has actually been instrumental in normalizing relations between Russia and China; many border crossings that were closed in the 1960s and 1970s were recently re-opened when China started allowing barter tourism with its Russian and Kazakh neighbors (Gormsen 1995; Nin 1994). Both sides have now opened up land crossings to international tourists in areas that were until recently forbidden.

Cross-border shopping is certainly not limited to international boundaries. Cities with higher sales tax rates than adjacent municipalities have always faced the problem of outshopping (Mikesell 1970). As well, in federal systems where individual states or provinces possess a great deal of sovereignty regarding economic matters (i.e. setting tax rates), such shopping patterns are common. While the primary motive for buying internationally is favorable exchange rates, for shopping in a neighboring subnational jurisdiction the motive is lower taxes (Clark 1994; Fox 1986; Mikesell 1971). In an early study of this problem, McAllister (1961) found that a lack of sales tax in Oregon and Idaho was the primary attraction for Washington residents shopping in the two states – an activity that increased dramatically when a 3.33 percent sales tax was introduced in Washington. Similarly, with no sales tax in Alberta, Saskatchewan, residents of Lloydminster, a medium-sized town split by the interprovincial boundary, have long done most of their shopping on the Alberta side of town to avoid their own province's sales tax. The problem got so bad at one point that local police instituted spot checks of shoppers along the boundary to find people guilty of tax evasion. Acting under provincial orders, police were required to enforce Saskatchewan sales and tobacco tax regulations to that province's residents shopping on the tax-free side of town. One Edmonton newspaper even headlined an article

'Lloydminster Fears a "Berlin Wall" in Crackdown on Sales Tax Evasion' (quoted in Dykstra and Ironside 1972: 271). As mentioned earlier, the federal government of Canada introduced a new seven percent Goods and Service Tax (GST) in 1991 throughout the entire country, which is attached to most non-essential products. Saskatchewan's provincial sales tax stands at nine percent, which, together with the GST, equals 16 percent tax. In Alberta the only sales tax charged is the GST, so that the same problem that existed 27 years ago still plagues the Saskatchewan side of town. While much of this local outshopping may not be considered tourism according to most definitions, it might, in some cases, involve recreational motivations, and even if it is motivated purely by necessity, residents still cross borders temporarily and make an economic contribution to the host jurisdiction.

Tourism of vice

It has been argued that when people are outside their home environments, they feel a sense of freedom from the 'puritanical bonds of normal living, anonymity is assured away from home, and money is available to spend hedonistically' (Mathieson and Wall 1982: 149). These forces, in conjunction with geographical advantages of the destination, contribute to the development of tourist activities that are widely considered to be negative, or immoral, such as gambling, prostitution, and drinking. Such activities are common in border communities throughout the world, either resulting from a planned effort or through the uncontrolled process of development. Gambling is a good example of planned development that fosters economic growth in many communities that draw visitors from jurisdictions where gaming is not permitted. Prostitution and drinking, however, are often the byproduct of a community's borderlands location, which, for the same reason as gambling, becomes attractive to certain groups of tourists.

Gambling

While gambling is widely viewed as an immoral activity by many community groups owing to its reputation as an evil that erodes traditional family values, devalues work, links communities to organized crime and political corruption, and creates social costs when it leads to compulsive behavior (Eadington 1996), it has received widespread acceptance as a tool for economic development. In fact it is promoted as the primary form of tourism in some destinations (Eadington 1996; Randerson 1994). While some people believe that once an activity is legal it is no longer immoral, others view gambling as a pariah industry whether or not it is legal. For purposes of tradition, it is included in this discussion on vices.

Like shopping, gambling tends to develop at border crossings when one polity that allows such activities meets another where gambling is prohibited.

The spatial development of gaming tourism, however, is generally not a random incident; borders are commonly targeted as strategic locations for deliberate development. The convergence of two systems provides opportunities for communities within the gambling state to attract large numbers of people from nearby out-of-state areas by establishing casinos and other gaming halls just inside their boundaries. So specific are the border dimensions that parking lots are sometimes bisected by state lines, and casino front doors lie within one or two meters of the border. Numbers of people flocking to adjacent states to purchase lottery tickets have grown immensely during the past decade (Adams 1995) often from convenience stores located just over the border. On a subnational level, a number of states in the United States have built casinos just inside their territories to cater to out-of-state visitors (Bowman 1994; Jackson and Hudman 1987; Weaver 1966).

With the development of the interstate highway system and with the increase in long-distance automobile travel, people in neighboring states and countries where gambling is not allowed are ready and willing to drive to the nearest border town for recreational purposes (Jackson and Hudman 1987). Together with this increased mobility, strict gaming policies in neighboring states, and legalized gambling in Nevada have spurred the growth of several small border towns in that state into thriving 'dens of iniquity' (La Ganga 1995; Sommers and Lounsbury 1991). 'The Nevada borderlands have, like a magnet, drawn new casinos to lure in gamblers as soon as they cross the state line' (Bowman 1994: 52). In that state's border towns alone, gaming by residents of other states amounted to more than US$55 million in 1991 (Sommers and Lounsbury 1991), of which nearly $1 million dollars per week were generated by residents of Utah, where all forms of gambling are strictly forbidden.

The importance of this phenomenon on an international level cannot be overstated. There are countless examples of one country attracting people from adjacent countries to participate in gaming activities. Monaco, which is perhaps one of the best examples of this, has long attracted tourists from France (residents and foreign visitors) (Leiper 1989). Like Monaco, Macau's economy has also been dependent on Hong Kong gamblers who regularly visit the Portuguese colony to play at its casinos and betting operations. This provides an important outlet for day trippers from a crowded city where casinos were not permitted (Leiper 1989; Lintner 1991b). Similarly, on St Martin, casinos have been established on the Dutch side of the island, which lure tourists and residents from the French side, and the casino in Taba, Egypt, relies on more than 90 percent Israeli patronage (Felsenstein and Freeman 1998).

In an effort to recover some of the money lost to outshopping by Canadians and to provide employment and tax dollars in economically depressed border towns, strict Ontario anti-gaming laws were relaxed in 1994 to allow the development of casinos adjacent to popular US border crossings in important gateway and tourist communities like Windsor and Niagara Falls (Smith and Hinch 1996; Truehart 1996). Casino Windsor does not attempt to hide the fact that its primary goal is to attract Americans across the bridge, and the company has spent large sums of money on advertising campaigns on the US side of the border (e.g. *Detroit News* 1997). These promotional efforts, close proximity to a large cross-border urban market (i.e. Detroit, Ann Arbor, Lansing, and Toledo), and favorable exchange rates for Americans have resulted in a successful Canadian business that is supported overwhelmingly (approximately 80 percent) by residents from abroad (*Travel Weekly* 1998). The popularity of Canadian frontier casinos is increasing pressure on their US urban counterparts to respond by opening up comparable establishments on their side of the border (Eadington 1996; Nieves 1996).

A similar situation existed in the Republic of South Africa (RSA) prior to political reforms in 1994. As part of the country's apartheid policy, the black population was segregated into homelands, unable to determine their own socio-economic and political status. In 1976, the RSA granted independence to the Transkei homeland. Bophuthatswana gained independence in 1977, followed by Venda in 1979 and Ciskei in 1981. According to critics, the gesture of granting nominal independence to the homelands was merely an effort to control the movement of Africans and to provide a labor reserve and dumping ground for the white population (Egerö 1991; Geldenhuys 1990; Merrett 1984). For this reason, the United Nations and its member countries never recognized the homelands as independent states; they were acknowledged only by South Africa and each other. Tourism developers and homeland governments quickly took advantage of this situation by establishing several resort complexes, including large casinos. By doing this, the homelands satisfied the demand for a tourist product that was illegal in RSA, but within easy reach of the country's major population centers (Stern 1987) – an activity that Lesotho and Swaziland had engaged in since their own independence from Great Britain in the 1960s (Crush and Wellings 1983; 1987; Harrison 1992). The homeland developments were designed to compete with existing casino resorts in Botswana, Lesotho, and Swaziland (Rogerson 1990; Southall 1983), and eventually Sun International established casinos in all four 'independent' homelands (Cowly and Lemmon 1986; Rogerson 1990). The most popular of these casinos, Sun City (also known informally as Sin City), located in Bophuthatswana, was drawing over one million visitors a year within a few years of its completion (Cowly and Lemmon 1986). With the recent re-integration of the homelands into RSA, the casinos have been allowed to continue functioning, and legislative action has been taken that now permits gaming activities within the borders of South Africa (Ahmed *et al.* 1999).

Some places are thriving on gambling that is permitted outside their territorial water limits. Although Israel has a national lottery, casino gaming is not yet legal. The southern port of Eilat is a popular destination for Israelis and foreigners alike, and part of the city's offerings include gambling cruises. Five boats operate out of Eilat, picking up passengers at the port, paying port exit fees for them as though they were leaving the country, and then sailing beyond Israeli territorial waters on four-hour gambling cruises (Felsenstein and Freeman 1998). Correspondingly, extra-territorial cruises are offered in communities along the southeast coast of the United States, and a similar concept is true along the Mississippi River in that casinos are permitted to function on riverboats in the water, but not on shore (Truitt 1996).

During the past decade, federal Indian reservation gambling in the United States has rocked the traditional gaming industry that has overwhelmingly been attracted to well-known centers such as Las Vegas and Atlantic City. These special administrative regions are under federal jurisdiction, not the state in which they are located. In 1988 the US Congress passed the Indian Gaming Regulatory Act (IGRA), a statute that governs the operation of Native American gambling establishments. The primary purpose of the IGRA was to 'balance the Indians' interest in tribal sovereignty with the state's interest in guarding its citizens from corrupt gaming activities and organized crime infiltration' (Greene 1996: 93). Following the enactment of the IGRA, dozens of court decisions favored tribal endeavors and resulted in a climate in which high-stakes gaming on reservation land flourished nationwide (Swanson 1992). As of 1995, Utah and Hawaii were the only states remaining with no permitted forms of commercial gaming, although 25 states had by then adopted casinos in one form or another (Eadington 1996). Between 1988 and 1995, the number of Indian reservations offering some form of gaming swelled to over 150 (Greene 1996), and reservation earnings in 1994 exceeded US$2.8 billion, up 40 percent from the year before (Eadington 1996; Gabe *et al.* 1996). In Minnesota alone, 16 Indian casinos were opened between 1990 and 1993, employing 10,353 people, 29 percent of whom were Native Americans (Gabe *et al.* 1996: 81). The IGRA permits the same types of gambling allowed anywhere else in the state for any purpose by any person, entity, or organization to be introduced within reservation boundaries (Eadington 1990; Swanson 1992). In other words, even if full-fledged casinos are not legal in a state, tribes are permitted to erect them if similar types of gambling are allowed occasionally to non-profit groups for fundraising purposes, which is a common practice in many states. Such developments on Native American land were viewed as a means of economic development that would create employment and attract capital investment (Stansfield 1996: 133). Gambling would also make up for cuts in federal funding by improving education, paving roads, building sewer and water projects, and funding rehabilitation programs for chemical dependency (Greene 1996).

Research by Lew (1996) demonstrates that tribal gambling occurs more often and is more likely on reservations in the eastern and mid-western US. Generally speaking tribes in the east are more aggressive in developing tourism compared to those in the west. Lew suggests that reservation size may have something to do with this. Larger reservations, he suggests, enable a greater sense of autonomy than do smaller ones. This means that residents of larger reservations, such as those in the west, are less likely to adapt to the dominant culture, including gaming tourism, than those in the East (Lew 1996: 361). For the smaller, eastern tribes who possess fewer tourism resources, gambling has become one of the most dominant forms of tourism in recent years; in western reservations most tourism still focuses on Native American culture.

One particular example of Indian gaming deserves mention in this discussion. The opening of Foxwoods Casino in February 1992 on the Mashantucket Pequot tribe's small reservation (726 hectares) was fraught with much state opposition. Nevertheless, the tribe gained permission from the federal government to open a high-stakes casino, because blackjack, craps, poker, and roulette were allowed in Connecticut at charitable 'Las Vegas nights,' which according to the IGRA gave that state's Native Americans free range with regard to casino gambling (Carmichael *et al.* 1996). In 1996, 23,178 square meters of gaming space, a capacity of 43,000 people, 4,428 slot machines, 308 table games, 21 food outlets, a theater, an atrium lounge, dozens of rides, an arcade, 1,500 hotel rooms, 7,721 parking spaces, a 3,000-seat bingo area, and 10,162 employees comprised this, the world's largest casino (Bixby 1996; 1997; Glassner 1996), which is still in the expansion process. Foxwoods brings in over a billion dollars a year – a sizeable sum for a community of only about 385 members (Bixby 1996; 1997). The establishment enjoys an excellent market location with about 22 million people, including Boston and New York City, within a 250-kilometer radius (Stansfield 1996). Acknowledging the success of the Pequots, several other east-coast tribes began plans to develop large-scale casino operations, and in October 1996, the Mohegan Tribe opened its casino just a dozen or so kilometers down the road from Foxwoods. The Mohegan Sun, the country's 132nd Native American casino and third largest in the world, is now in competition with Foxwoods for gaming dollars (Bixby 1996).

Prostitution and pornography

Prostitution often follows large-scale tourism development, and border regions are particularly vulnerable to the growth of sex tourism. Two primary conditions work together that might explain the growth of border prostitution. First, although many frontier regions are becoming heavily industrialized, such as the US–Mexico border and in Europe along the old Iron Curtain (*Economist* 1993b), they still are unable to provide sufficient employment for the masses of people who flock there from the interior of the country, so that many people are driven into menial service occupations, including prostitution. Fernandez describes the situation as follows:

> The border towns have become famous among people throughout Mexico for having the highest wage scales in the entire Republic. As a result, an increasing number of peasants in the interior are making the decision to sell their homes, their cattle, and whatever other belongings they may have, in order to migrate to the border area. They go there expecting to find a wonderful job awaiting them. But a very large percentage are disappointed to discover upon arrival the true state of affairs in border towns. The female is fortunate who finds a job in a factory; rarely does a male find one. Great numbers of unemployed peasants are thus stranded in the border towns. Some return to the interior. Most do not. They have nothing to return to – all of their possessions have been sold.
>
> (Fernandez 1977: 129)

Second, in common with gaming, borders are ideal locations for prostitution because visitors from a neighboring country do not have to venture far into foreign territory to find their pleasure outlets. Large population centers near the border on the side that does not permit such activities provides a substantial market base for these illicit activities.

During the prohibition era (1918–33), when the manufacture, sale, and transportation of alcoholic beverages were banned from the United States, Mexican border towns emerged as important playgrounds for drinking, gambling, and prostitution for Americans. The heyday was short-lived, however, as border tourism collapsed during the Depression of the 1930s, when prohibition was recalled, and many of the houses of ill-repute were closed. After the Second World War, business again picked up and continued to grow through the 1950s and 1960s, especially in border towns located near US military installations where servicemen sought 'to provide for relief to their sexual needs' (Fernandez 1977: 127) and where US border residents could purchase rationed commodities such as gasoline, tires, groceries, and metal car parts (Arreola and Curtis 1993). Thus the Mexican frontier towns emerged as 'convenient yet foreign playgrounds, tantalizingly near but beyond the prevailing morality and rule of law north of the border' (Curtis and Arreola 1991: 340), and they gained a reputation as 'havens of hedonism' (Arreola and Curtis 1993; Curtis and Arreola 1991). By the 1970s, prostitution along Mexico's northern border had become such a ubiquitous and popular activity, that guides were even published to direct American tourists (mostly men) to hot spots all along the border where their every uninhibited fantasy could be realized (e.g. West 1973). In telling about the red light district in Nuevo Loredo, Mexico, one publication suggested:

> The one entrance and exit is carefully supervised by Nuevo Loredo's finest. Girls come from all over Mexico to work in the four or five nicest places. This is the best on the border ... Most unique and pleasing

> structurally, both in building and clientele, is the *Tamyko*, a huge Japanese pagoda with outside patio and fishponds spanned by arched bridges, surrounded on two levels by bedrooms. The girls are almost all between 14 and 18 years old.
>
> (West 1973: 73)

Prostitution, which is legal in Mexico, does not exist today to the extent that it did between the Second World War and the 1970s. The past two decades have brought about a reorientation in economic activities in the US–Mexico borderlands. With the growth of the *maquiladora* industry along the Mexican side of the frontier, many of the people who previously worked as prostitutes have become productive workers in the US-owned assembly plants. It appears that there are fewer prostitutes and the activity plays a far less significant role in the border economy than in the past. In addition to the industrialization of the borderlands, other changes are at play. In the USA the emergence of topless and bottomless bars, increased sexual permissiveness, increased availability of pornographic movies and videos, the spread of pornography shops, massage parlors, and escort services, and the fear of AIDS, have decreased the competitive advantage of Mexican border towns (Curtis and Arreola 1991: 343). No longer does the border possess the magnetic effect that it once did for sex tourism. Nevertheless, the 'industry' still draws crowds of Americans owing to lower prices and an exotic setting (Bowman 1994), but most clients are residents of Mexico (Curtis and Arreola 1991: 337). Similar situations developed in Lesotho and Swaziland in conjunction with the gaming activities discussed earlier. Pornographic films that were banned in South Africa were a hit among that country's residents visiting Lesotho and Swaziland, and more informally, 'prostitution across the color line became the cornerstone of Lesotho's and Swaziland's tourist attractions' (Crush and Wellings 1987: 101).

Increased tolerance of vices in the United States has produced a similar situation on the US–Canada border to the one that existed earlier in Mexico. During the 1970s, at the time the run for the Mexican border began to decrease, the run for the US border by Canadians began to increase. Some US border communities became known for their more liberal drinking, gambling, and pornography laws than those in Canada. Blaine, Washington, for example, an otherwise peaceful and conservative small town, became known as 'Tijuana of the North' (Simmons and Turbeville 1984). The town offers drinking, gambling, and pornography seven days a week, and it became a well-known destination for residents of nearby crowded Vancouver (Bradbury and Turbeville 1997). It also became a frequent side trip for visitors to Vancouver from other parts of Canada, because British Columbia's restrictive drinking and censorship laws created the need for outlets for this type of behavior. Because Canadian laws required severe editing of pornographic films, books, magazines, videos, and peep shows, Blaine developed into a prosperous porno paradise 'normally found only in much larger cities' (Simmons and Turbeville 1984: 52–3). Recent growth in the video business and the gradual liberalization of strict controls in British Columbia have begun to decrease this genre of tourism in Blaine.

Drinking and other vices

In common with gambling and prostitution, borders are fitting locations for drinking establishments. People flock across political boundaries to purchase liquor for home use or to spend time drinking in bars and pubs elsewhere. Several variables encourage this type of cross-boundary travel: cheaper alcoholic beverages; lower liquor taxes; lower drinking ages, if they exist at all; and longer opening hours.

Canadians can purchase Canadian beer in the US cheaper than in Canada, and even when import duties are paid on their return, the overall cost is still cheaper than at home. A drive through American border communities reveals how ubiquitous this phenomenon is as many markets, bars, liquor stores, and convenience shops erect special signs and offer bargains to potential Canadian drinkers. A pint of draft beer in Vancouver costs more than twice as much as in Point Roberts, Washington, and until 1986 the small community was the place closest to Vancouver where Canadians could buy beer on Sundays (McAllister 1996). A similar situation has long characterized much maritime travel between England and France (Matley 1976). These 'booze cruises,' as they were known, were popular among the British, who could travel by boat quickly and cheaply to French and Belgian ports where alcohol taxes were much lower than at home. Cheap drinks provided 24-hour entertainment, usually starting on the boat before it left the harbor in England (Hidalgo 1993). Matley (1977: 25) called this activity 'alcohol tourism' and argued that it was of questionable value, especially considering that it attracted the worst type of tourist, resulting in drunkenness, public disturbances, violence, and vandalism. With the elimination of border controls and changes in the tax structures within the European Union, this activity has slowed somewhat but is still going strong.

This issue becomes particularly problematic when underage youth pour across borders by the thousands in search of cheap drinks and bartenders who do not check for identification. All along the US–Mexico border, American teenagers as young as 14 regularly cross to drink, and until recently Tucson, Arizona, radio stations ran advertisements for Mexican clubs that targeted Arizona teens. The city's stations no longer accept ads from teen-oriented clubs south of the border, but the bars have circumvented this obstacle by paying students to hand out fliers at school (Bowman 1994: 62). Spring break is a particularly popular time to get away from the confines of home and indulge in unruly behavior. Many high school and university students plan cross-border trips hoping to find bars that remain open all night and an absence of a legal drinking age (DeQuine 1989). Padre Island, Texas, for example, thrives on spring break students, many of whom base their vacation decisions on the fact that the island is located near the Mexican border where beer and liquor can be purchased far cheaper than in Texas (Gerlach 1989: 16).

In addition to alcohol, gambling, and sex, Americans have historically traveled south in search of more exotic entertainment, such as bullfighting, cockfighting, and boxing – activities that were either illegal or too expensive in the United States (Arreola and Curtis 1993: 98). Drugs too are readily available in many border communities where some law enforcement officers turn a blind eye to these illicit activities.

International parks

There has been a recent growth in the number of international parks that lie across or adjacent to international boundaries. In borderlands where unique ecosystems and cultural heritage are protected, international parks have been established since the early 1900s. In most cases two, and sometimes three, parks meet at an international boundary, although some international parks occur on only one side of a border. Denisiuk *et al.* (1997) and Thorsell and Harrison (1990) identified over 70 borderlands parks and nature reserves throughout the world.

Czechoslovakia and Poland initiated the notion of cross-border cooperation for the development of frontier parks. The two countries signed an agreement in 1925 that would later assure the development of three international parks between 1948 and 1967. Likewise, bilateral legislative action in the United States and Canada in 1932 established Waterton-Glacier International Peace Park. This united Waterton Lakes National Park (Canada) with Glacier National Park (USA), with the goal of promoting peace between neighbors and nature conservation on both sides of the boundary. These early events laid the groundwork for the establishment of dozens more international parks during the twentieth century, and many more are currently in the process of being established (Timothy 2000b) (Figure 3.4).

Borderlands are ideal locations for parks and protected areas because of their attributes (e.g. peripheral, marginal, sparsely-populated, underdeveloped, and isolated). Young and Rabb (1992) reason that the old frontier zones of Eastern Europe contain some of Europe's most fascinating and undisturbed natural scenery and wildlife. The fact that so many of Central and Eastern Europe's national parks already exist along national boundaries attests to this fact (Denisiuk *et al.* 1997). Efforts are under way to create international parks in those regions, and according to Westing (1993), several hundred of the nearly 7,000 protected natural areas that existed in the mid-1990s throughout the world are either adjacent to, or in close proximity to, national boundaries.

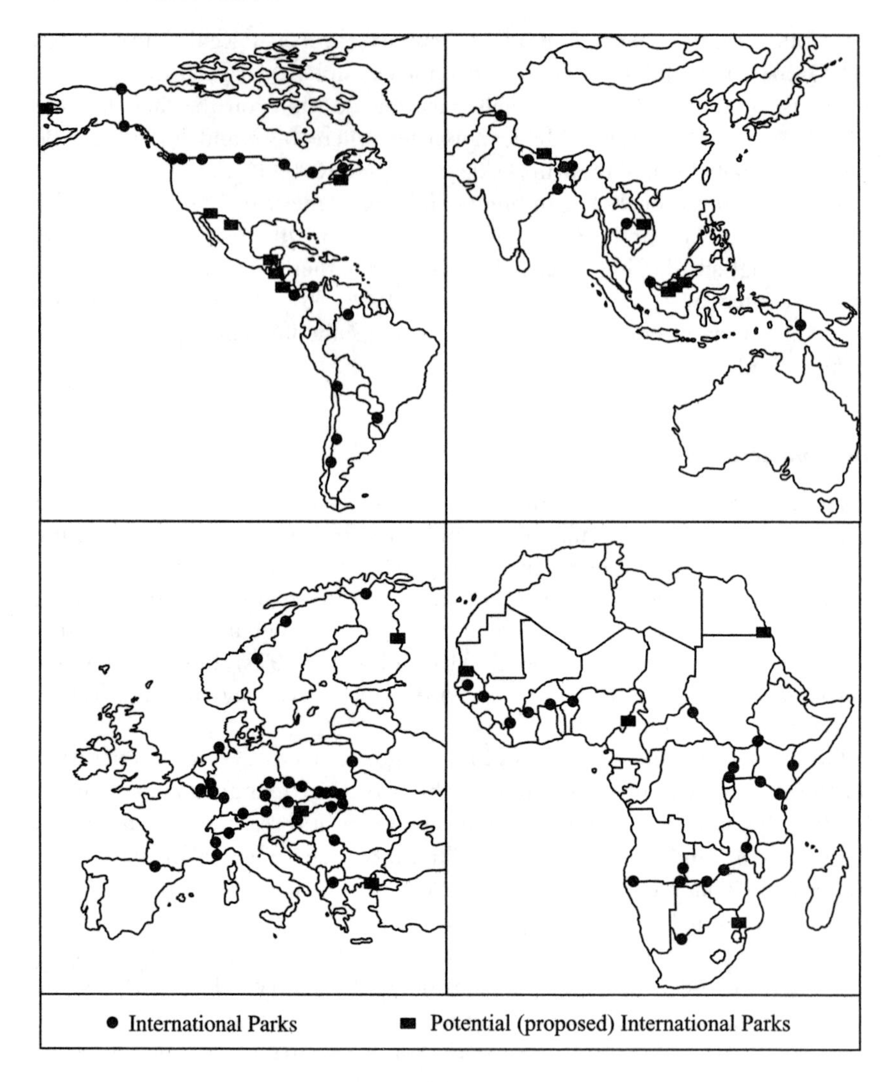

Figure 3.4 Locations of international parks
Source: Timothy 2000b

Just as national parks are one of the most essential types of tourist destination in the world, by extension, international parks have become important global destinations. Places like Waterton-Glacier, Iguazu Falls, Victoria Falls, and Masaai Mara-Serengeti attract millions of visitors every year from around the globe. Other lesser known international parks are more important destinations for domestic tourists and visitors from neighboring nations, or as secondary attractions for tourists who are already in an area for other primary purposes (Timothy 2000b).

To demonstrate the scale of tourism, Table 3.3 shows annual visitor numbers from three international parks in North America. Waterton-Glacier International Peace Park is an important global tourist destination located in the US and Canadian west. It attracts approximately two million visitors a year to the American side and nearly 400,000 to the Canadian side each year. More than 200,000 people visit the International Peace Garden each year, primarily from the US mid west and Canadian prairie provinces. The Garden is popular among young people for its music and sports camps, which bring in youth from many countries each summer. Visitor numbers at Roosevelt Campobello International Park range from 110,000 to 150,000 annually. Most visitors come from the eastern United States and Canada, although some come from other parts of North America and overseas (Timothy 1999a).

In situations where binational relations are good, crossing from one side of a park to the other can be done relatively easily. This is so in most cases and allows both sides to benefit economically as profits are spread to communities on both sides of the boundary. It also allows visitors to view the local ecosystem more holistically. For some individuals, seeing the attraction from both sides of the boundary, and even experiencing the border itself within the park

Table 3.3 Visitor numbers to international parks in North America, 1990–97

Year	*Number of visitors*	*Year*	*Number of visitors*
Waterton-Glacier International Peace Park			
	Waterton		Glacier
1990	353,908	1990	1,987,000
1991	344,028	1991	2,096,966
1992	345,662	1992	2,199,767
1993	344,453	1993	2,141,704
1994	389,510	1994	2,152,989
1995	364,740	1995	1,839,518
1996	346,574	1996	1,720,576
1997	370,733	1997	1,708,877
International Peace Garden			
1990	197,327	1994	222,093
1991	203,247	1995	218,036
1992	209,345	1996	211,495
1993	215,625	1997	205,151
Roosevelt Campobello International Park			
1990	144,678	1994	132,551
1991	151,327	1995	122,682
1992	138,950	1996	111,431
1993	135,842	1997	121,530

Source: Timothy 2000b

can be a highlight of the visit. In most instances passage between sides requires frontier formalities, but in some cases, border crossing goes unhindered and unregulated within the parks (e.g. Peace Arch Park and the International Peace Garden).

International exclaves

International exclaves, or enclaves, are another important tourism destination in some regions, particularly in Western Europe. These unique communities, which have heretofore been overlooked by tourism researchers, are almost all dependent on tourism for their economic well-being. True exclaves are small parts of one country entirely surrounded by a neighboring country (Robinson 1959) (Figure 3.5). From the perspective of the country to which it belongs, the home state, it is an exclave. From the perspective of the country in which it is located, the host state, it is an enclave. Most exclaves are located in Europe and Asia, but a few inhabited ones are also located in North America (see Table 3.4); Switzerland hosts two (Campione, Italy and Büsingen, Germany); France surrounds Llivia, Spain; the Netherlands hosts 22 Belgian enclaves (Baarle-Hertog); and Belgium is home to at least 8 Dutch outliers (Baarle-Nassau), some of which lie within the Belgian exclaves themselves. One Omani exclave lies within the United Arab Emirates. India and Bangladesh both own numerous small tracts of territory within the other country along their common border; and with the collapse of the Soviet Union, 13 new exclaves were created (Catudal 1979; Timothy and

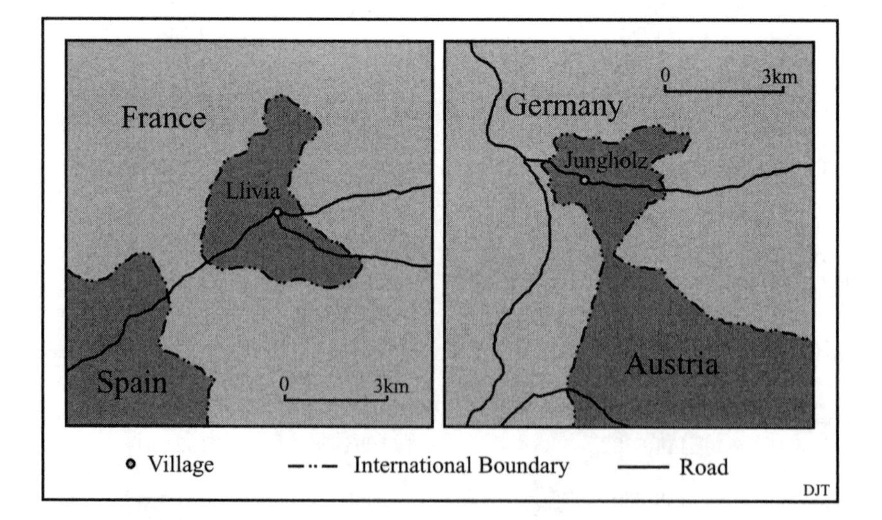

Figure 3.5 Examples of an exclave (Llivia, Spain) and pene-exclave (Jungholz, Austria)

Table 3.4 Tourism-oriented exclaves in Europe and North America

Exclave	*Enclaving country*	*Size in area (km^2)*	*Approximate population*	*Primary attractions*
Baarle Hertog (Belgium)	Netherlands	7.25	8,000	Smallness, political situation, market, built heritage
Büsingen (Germany)	Switzerland	7.62	900	Smallness, political situation, historic church, restaurants
Campione (Italy)	Switzerland	2.60	2,400	Smallness, political situation, gambling, built heritage, art
Jungholz (Austria)	Germany	7.80	340	Smallness, political situation, skiing, mountain hiking
Llivia (Spain)	France	12.87	900	Smallness, political situation, skiing, spas, festivals, museums
Northwest Angle (USA)	Canada	337.00	75	Smallness, political situation, fishing, hunting, winter sports
Point Roberts (USA)	Canada	10.60	300	Smallness, political situation, shopping, summer homes, boating

Source: Catudal 1979; Timothy 1996

Mao 1992), although their present status is uncertain. Likewise, the term pene-exclave is used to describe small outliers that are physically connected to the homeland, but owing to physical barriers, are easily accessible by wheeled traffic only by traversing foreign territory (Robinson 1959). Pene-exclaves are located along the Germany–Austria (e.g. Kleinwalsertal and Jungholz), Ireland–Northern Ireland (e.g. Drumully), and Italy–France (e.g. Bagni di Craveggia) borders. In North America several are located along the US–Canada border, but only a few are inhabited: Point Roberts, Washington; Northwest Angle, Minnesota; a portion of the St Regis Indian Reserve, Quebec; and Campobello Island, New Brunswick.

Some exclaves are remnants of medieval property ownership by nobility and ancient communal jurisdictions (e.g. the Baarles, Campione, Büsingen, and Llivia). Others are the results of misunderstandings during boundary delimitation and demarcation or of just being in the way of a straight-line boundary (e.g. Northwest Angle, Point Roberts, St Regis, and Campobello Island), and until 1990 West Berlin and Steinstücken were examples of exclaves created by military occupation. These entities also sometimes undergo the process of disenclavement, wherein they cease to exist because they are either traded with the host state for equal portions of territory, given freely, absorbed by the host state, or disintegrated when home and host states are unified (Catudal 1979). Most European and North American enclaves are populated with fewer than 2,500 residents. The exception is Baarle's population of about 7,900. In addition, most range in size from 2.6 square kilometers (Campione) to 83 square kilometers (Northwest Angle) and most lie only one or two kilometers from the frontier of the home country.

Travelers tend to be fascinated by smallness, which, according to Jenner and Smith (1993), is one of the primary reasons why millions of tourists are attracted to the smallest countries of Europe every year. Similarly, tourists tend to be intrigued by the smallness of exclaves and their unusual geopolitical situations (Timothy 1996; Timothy and Mao 1992). The legal status of enclaves, territory within a country but outside its legal control, sometimes creates additional elements of attractiveness. With the exception of Point Roberts and Campobello Island, most exclaves traditionally have had open borders without customs and immigration controls (Plate 3.8). In common with other borderlands situations, where gambling, drinking, and pornography laws differ between home and host state, exclaves belonging to the more tolerant country attract visitors from the enclaving country where such activities are not permitted (Minghi and Rumley 1972). For example, as mentioned earlier, prohibitive drinking laws in British Columbia, Canada, for many years have pushed travelers to Point Roberts and other American border towns where liquor laws are not as restrictive (German 1984; Minghi and Rumley 1972). Moreover, shopping is popular in some outliers owing to lower prices and tax rates than in the host country, and a different range of goods is usually available. For these attractions, exclaves are popular because travelers can enjoy the benefits of being in a foreign country without having to pass through typical border-crossing formalities (Timothy 1996).

In addition to its attractive mountain setting, Llivia draws a significant crowd of tourists to its shops, health spa, and fourteenth-century buildings (Llivia Municipal Museum Trust 1986). Tourist services include a tourist information office, four hotels, and seven restaurants. As well, the village hosts the annual Villa de Llivia Music Festival and functions as a base for winter tourists who can access over 15 ski resorts within a 45 kilometer radius. Recent data are not available, but it appears that in the past most tourists to Llivia originate in France (Catudal 1979). As mentioned earlier, a complex mesh of small pieces of Belgium, some no larger than a single building, intermixed with several tiny bits of Holland, some also no bigger than a single

Plate 3.8 The entrance gate on Campione d'Italia's border with Switzerland leads to a variety of gaming, shopping, and heritage opportunities. There are no immigration or customs formalities at this crossing

house and yard, constitute the town of Baarle. This abnormal geopolitical situation is Baarle's most important attraction for tourists, but the town is also known for its colorful market place and historic buildings, some of which are even more intriguing because they are bisected by the international boundary (Baarle-Nassau Tourist Office n.d.). The community has a tourist information office, several hotels, and numerous dining establishments. One estimate from the 1970s suggested that 40,000–50,000 tourists visited the Baarles each month, and that approximately 35 percent of the local population were involved in service industries (Catudal 1979). According to one official at the tourist office in 1995, the number of visitors has increased substantially since then. Precise numbers are not available, but officials estimate that between 60 and 80 percent of Baarle's visitors are Dutch. The rest are from Belgium and other European countries. Campione d'Italia is best known for gaming and live entertainment at its Casino Municipale. However, tourism developers are beginning to emphasize the community's art works and architectural heritage in their promotional campaigns, and the community recently hosted a world-class speed boat competition along its shore in Lake Lugano (Azienda Turistica 1994). In addition to the tourist information office, services consist of numerous hotels and restaurants. Catudal (1979) estimated that tourism and related service industries comprised 90 percent of the workforce in Campione in the 1970s, a figure which was confirmed by the local tourist office for 1995 as well. According to officials, the majority of tourists are Swiss. Jungholz, Austria, which is only accessible via Germany, has developed into a rather impressive ski resort. Summer visitors are also drawn to the extensive system of hiking and mountain bike trails (Touristikinformation Jungholz 1994). The local tourism infrastructure includes an information office, a campground, a ski school, and several hotels, guesthouses, and restaurants. The majority of visitors are from Germany. Büsingen is the least tourist-oriented of the European exclaves, although it is popular among Swiss visitors from nearby Schaffhausen for its fine dining. The town's only real attraction is its old Bergkirche church, and the local tourism infrastructure is currently limited to one hotel and five restaurants. In the 1970s the most significant economic activity was agriculture, which employed nearly 80 percent of the population. Services, including tourism and self employment, accounted for the remaining 20 percent (Catudal 1979). According to the mayor's office in 1994, those estimates are still quite accurate, except that agriculture is less important today since many of Büsingen's residents work in Schaffhausen, Switzerland.

In the North American context, shopping is the primary attraction in Point Roberts for Canadians. Although the town lacks major shopping facilities, sales of gasoline and groceries were an important part of the local economy during the late 1980s and early 1990s, as it was in many other US border communities (Timothy and Butler 1995). In addition, Point Roberts boasts a sizeable marina and is a popular second-home resort, where most summer residents are Canadians (German 1984; McAllister 1996). Tourism in the Northwest Angle is based on fishing and hunting in summer and cross-country skiing and snowmobiling in winter. The Angle has 13 hunting and fishing lodges, one pub, several dining halls, a country club and golf course, an air strip, and a trading post. Furthermore, several festivals have been organized in recent years and together with the recently restored Fort St Charles, which dates from 1732, add a cultural heritage dimension to the community's attraction base (Timothy 1998d). Estimates by Angle lodge owners place tourist numbers at approximately 8,000 per year on average – a significant number in a community whose permanent population is comprised of only about 70 people. It is also estimated that tourism employs more than 90 percent of the population and that 96 percent of the tourists are from the United States, while only four percent are from Canada (Timothy 1998d).

Timothy (1996) examined four unique obstacles to tourism development in exclaves not commonly confronted by most destinations. First, many enclaves are physically isolated in the sense that they are located in topographically difficult regions. This is especially true in the cases of Northwest Angle, Llivia, and Campione. Second, political accessibility is a problem for all enclaves. Treaties are usually signed between home and host state that permit the free flow of goods, security personnel, and residents between the enclave and the home country through host territory. Although foreign states have no official jurisdiction over enclaved territories, they sometimes place restrictions on the flow of certain items through their own territory. For example, American fishers are not permitted to carry live bait through Canada, even though it is for their exclusive use in the Northwest Angle. Similarly, host countries can blockade enclave borders to inhibit the growth of activities that are viewed as undesirable. For instance, when a casino was built in Campione in 1933, gambling was legal in Italy but not in Switzerland. The Swiss responded by blockading the entire enclave, permitting no visitors to enter until Italian officials agreed to a compromise that restricted the gambling activities of Swiss citizens (Pedreschi 1957). Third, small size restricts the physical development of tourism infrastructure. Large projects, such as transportation centers, are rarely appropriate, since the boundary which surrounds enclaves necessarily limits the spatial expansion of tourism. Smallness also limits natural and cultural resources upon which tourism can be based. Another drawback of smallness, but different from the situation of small independent states such as Andorra, Liechtenstein, San Marino, and Monaco, is that exclaves are not sovereign and therefore are not free to establish their own tourism policies and form international alliances for development as Liechtenstein has done with Switzerland and Monaco with France. Finally, something related to small size and political accessibility, is the question of the provision of public services. Smallness tends to preclude the development of electric, sewage treatment, and water treatment plants large enough to meet the needs of residents and tourists. As well, political boundaries usually impede the diversion of water, electricity, and gas from home state through host territory. This condition can be extremely problematic if the host country refuses to provide services and refuses to allow the provision of services by the home state through its territory. Agreements are usually made regarding which country will provide such services, but variations occur from place to place. The Northwest Angle receives its electricity from the Canadian province of Manitoba. Llivia's source of water and electricity is Spain. Büsingen's postal and telephone service is German operated, but Switzerland provides the exclave with gas, water, electricity, and transportation. In Campione, Swiss authorities operate the telephone system and post office, although Italian stamps are used, and utilities are provided by Switzerland. In the Baarles, water and electricity are supplied by the Netherlands, while each country operates a post office in its respective part of town, and Dutch and Belgian buses circulate throughout the community (Timothy 1996).

The mini microstates

In most of the geo-political and development literature, the term microstate refers to small nations with populations under one million and includes islands and non-island states (Wilkinson 1989). While this definition includes such physically large continental states as Botswana, Oman, and Guyana, most tourism research on microstates has focused on small islands (e.g. Baldacchino 1993; Fagence 1997; Milne 1992; Royle 1997). The purpose of this discussion is not to examine island microstates or physically large states with small populations. Rather, in keeping with the theme of political boundaries, this section aims to discuss tourism in the smallest of the microstates in terms of both population and land area, and the European states (i.e. Andorra, Liechtenstein, Monaco, San Marino, and Vatican City) will be used as examples. Since most of these countries range in size from 468 square kilometers (Andorra) to 0.44 square kilometers (Vatican City), tourism is indeed a border phenomenon. Even if visitors are drawn to the center of these countries they are still in effect in the borderlands.

In all five of Europe's smallest countries, tourism is an important economic sector. This is reflected in the high levels of touristic development, the nature of the attraction base, and the number of people employed in the industry. Shopping and skiing are the primary attractions offered in Andorra (Reichart 1988; Sanguin 1991; Taillefer 1991). Andorra typically attracts over 12 million visitors a year (*Economist* 1993a; Jenner and Smith 1993), although most of these are day trippers who shop, see a few sites, and then head back home or on to their next destination (Reichart 1988; Rinschede 1977). Nonetheless, Andorra is different from the other microstates in this respect in that it has approximately six times the number of overnight visitors of the other four countries combined and the largest number of hotel rooms (11,571 in 1991) (Jenner and Smith 1993). Of the 54,507 population in 1990, 35.5 percent of the workforce was employed in tourism.

Liechtenstein's primary attraction is its mountain setting wedged between Austria and Switzerland. Skiing is an important tourist activity as are hiking, historic sites and museums, and shopping (Liechtenstein National Tourist Office 1997). Approximately 60,000 overnight tourists visit the country every year and guest beds in 1995 totaled 1,290. Tourism places high in economic importance and nearly 14 percent of the population are employed directly or indirectly in tourism (Verwaltungs- und Privat-Bank 1998). While Monaco is best known for gambling, gardens and built heritage also form an important part of the attraction base (Direction du Tourisme 1994). Monaco, located on the Mediterranean coast, possesses a strong tourism infrastructure, including 2,350 hotel rooms in 1991. Roughly 13 percent of the population is employed directly in the tourism industry (Jenner and Smith 1993). San Marino, known as the world's oldest republic, is a unique destination located entirely within Italy. Its attractions include architecture, history, scenic landscapes, postage stamps, art, crossbow tournaments, and museums, as well as its smallness (Cohen 1993; Jenkins 1988; Mann 1987). Approximately 17 percent of the workforce are employed in tourism directly and the country had 545 hotel rooms in 1991 (Jenner and Smith 1993). While precise numbers are unknown, it is likely that nearly all tourists who visit Rome also visit the Vatican. If this is the case, then tourist arrivals in this, the smallest nation in the world in population and area, far exceeds the other European microstates. Public accommodation is not provided since tourists are not permitted to stay overnight in the country unless they are visiting friends and relatives. The percentage of the workforce in tourism is not known, but is believed to be relatively small since Catholic Church activities dominate. In these ways, Vatican City is unique among the small states of Europe.

Although the smallest of the land-based microstates have many characteristics and problems in common with islands and other types of small nations, they also face unique challenges that can act as major attractions and obstacles to tourism development. In most cases, small size produces both advantages and disadvantages. According to Diggines (1985), the closeness and intimacy of a small society produces a feeling of identity of individuals with their whole community – something very difficult to achieve in larger nations. Smallness creates a fairly homogeneous population which assures more peaceful relations than in larger countries where ethnic groups are constantly at odds (Diggines 1985), a particularly meaningful consideration for the development of tourism. Butler (1993) suggested that the feeling of separateness, of being cut off from the mainland, and political independence is a significant psychological attribute of islands that appeals to tourists. The same can be said of the smallest microstates, since smallness manifests itself in the tourist psyche as an 'exotic quaintness' (Baldacchino 1994: 51). Rinschede (1977) attributed Andorra's unique political situation and small size to its successful development as an important tourist destination. According to Baum (1997: 23),

> Political independence and the attraction of 'going abroad' to a small, neighbouring country applies equally to visits to Liechtenstein and Andorra as it does to Malta. What the visitor is experiencing is the excitement of travelling across a political border ... and also the opportunity to enjoy different political, infrastructural, cultural, and, on occasions, linguistic experiences at a micro level.

Regardless of the apparent success of tourism in these tiny nations, several negative aspects of small size have also been identified by various authors (e.g. Diggines 1985; Wilkinson 1987). In common with the exclaves discussed earlier, the narrow limits of territory restrain spatial development. For example, several years ago when Andorran tourism managers decided to develop a gambling casino to draw people from nearby France and Spain, the plans failed because there simply was no building site big enough in the country to fit the ambitious project (Reichart 1988). A limited natural and cultural resource base means that tourism may become too overdependent on too few attractions and tourism types. Limited human resources also creates a situation wherein the labor force is likely to have a narrow spread of specialization, which could result in frequent and unpredictable shortages of certain skills (Baldacchino 1993: 34). In terms of economic impact, short length of stay is another tourism problem created by small size. According to Jenner and Smith (1993: 69), 'It is ... difficult ... to encourage tourists to stay long in countries you can drive all the way through in anything from a couple of hours to as little as a few minutes.'

Several other political and economic difficulties arise in small countries that profoundly affect tourism. Their political isolation makes them more vulnerable to outside forces that are beyond the control of local developers and planners (Baldacchino 1994) and tends to render them unable to influence international events that may closely affect their interests and unable to provide or afford adequate overseas representations (Diggines 1985; Lockhart 1997a). In fact, many of the microstates face a constant battle for international recognition and individuality (Cohen 1993).

Most small states have domestic economies and markets that are inadequate in size and too dependent to allow worthwhile economies of scale (Payne 1987; Royle 1997). This results in limited domestic opportunities for higher education, press, radio and television services, as well as an often inadequate infrastructure and high transportation costs (Diggines 1985; Lockhart 1997a).

Cultural and political day trips

While international, one-day excursions characterize most cross-border shopping, tourism of vice, and visits to exclaves and mini states, other types of international day trips also exist based on people's curiosity about the political and socio-cultural differences that lie across the border, and their own heritage. For example, with the collapse of the Soviet Union, many of the people who were born before the Second World War in the Finnish areas now occupied by Russia began returning to the region of their birth (Karelia) on organized day trips to see the communities, factories, and farmlands they had left behind over 50 years earlier. While most people who visit are deeply disappointed by the dramatic changes that have occurred to their home towns under Soviet rule, they appreciate the nostalgic opportunity to explore the lands of their childhoods (Paasi 1996; Paasi and Raivo 1998).

Every year thousands of older Americans migrate to the southernmost extremes of the United States to spend their winter vacations in warmer climates. Retirement communities near the US–Mexico frontier are particularly popular, because they are within easy range of day trips across the border. This added amenity enhances the attractiveness of the sunbelt for many older Americans, particularly in Arizona and the Rio Grande Valley of Texas (Wilson and Mather 1990).

Day trips to China from Hong Kong and Macau have long been popular among foreign tourists (Gormsen 1995), largely because of the mystique that westerners associate with the country owing to its place in the world as the largest remaining communist regime (Cao and Ge 1993; Storey 1992). For the same curious reason, Americans elude US government restrictions on travel to Cuba for regular vacations (Reynolds 1997), but others pay US$200–300 for single-day trips to the island from Montego Bay, Jamaica.

While complete freedom of movement is still out of the question, some Chinese are being allowed to travel to neighboring countries on a limited basis, as the barter tourism example discussed earlier demonstrates. Cross-border day tours to Myanmar and Laos were initiated in 1991. After many years of isolation, and owing to their curiosity about what lies outside their country, masses of Chinese citizens are beginning to take advantage of this new form of tourism. Aside from the fact of just being abroad, culture, shopping, and food are the most notable features of these tours (Chenming 1992).

Participants in these day trips, as well as cross-border shoppers, exclave and ministate visitors, and participants in gambling, prostitution, and drinking comprise one of the largest cohorts of international tourists in the world. In the mid and late 1980s it was estimated that over 740 million international day visits were made throughout the world to a value of over US$109 million (Bar-On 1988). According to Murphy (1985: 5), day trippers are an important type of tourist because they spend money and time while utilizing space and facilities in the destination area. Bar-On (1988) echoes this conclusion, and suggests that day trippers include cruise passengers as well as people who cross frontiers by other modes of transportation.

Other attractions

As mentioned earlier, very often border markers themselves are primary tourist attractions. However, many other types of tourist attractions that are not directly a consequence of the existence of a border are also present in border regions (Boyd 1999; Butler 1996; Paasi and Raivo 1998).

Several well-known beaches are bisected by international boundaries, including the eastern end of the Gulf of Aqaba, where a 20-km stretch of attractive seashore is divided between Egypt, Israel, and Jordan (*Economist* 1995). Historic sites also exist in several borderland locations (Boyd 1999). Preah Vihear, for example, is an ancient temple complex located just a few hundred meters inside Cambodia near the border with Thailand. It is one of the finest examples of Khmer architecture and one of the most impressive temples in Southeast Asia. It is also a significant tourist attraction, although visitation levels vary dramatically according to the intensity of political unrest in the region (St John 1994). Likewise, nature-based tourism is common in border regions, owing largely to their typically peripheral locations. Rain forests along the Costa Rica–Panama and Uganda–Democratic Republic of Congo borders are acclaimed areas for ecotourism and various other forms of nature travel (Timothy 2001). While these tourism types are not necessarily a result of their proximity to an international border, they are influenced by it.

Summary

The focus of this chapter is the situation where borderlands are tourist destinations. Because they mark differences in societies, economic systems, and political regimes, borders wield a sense of fascination among travelers. Throughout the world examples abound where border markers themselves have become considerable tourist attractions, and the heart of some communities' tourism industry is the existence of these cultural and political icons. This is certainly the case at Four Corners in the United States, where Utah, Arizona, Colorado and New Mexico touch. In other places, frontier theme attractions and non-political lines, such as museums and parks, and the Equator, respectively, capture the tourist gaze. Sometimes, international boundaries pass directly through renowned cultural and natural attractions like Niagara Falls, a situation that can have serious implications for the planning and cross-national management of the sites.

In addition to the line itself, the areas adjacent to boundaries quite commonly become significant destinations as activities like gambling, shopping, prostitution, and drinking come about. These have a tendency to develop adjacent to borders across from regions where comparable pastimes are not allowed to flourish. Likewise, international parks are intriguing to some visitors because they can experience the offerings from two vantage points. Most of the exclaves in Western Europe and North America and the smallest microstates of Europe have developed an economic dependency on tourism owing to their popularity as small yet foreign destinations where cultural heritage, natural landscapes, and smallness contribute to making them fascinating objects of tourist attention.

4 Landscapes of borders and tourism

> Niagara Falls, Ontario, may resemble a second-rate amusement park at times, but Niagara Falls, New York, resembles a second-rate amusement park that's been closed for twenty years.
>
> (Schwartz 1997: 26)

Introduction

The impact of human behavior on the physical environment has long been a primary research interest of human geographers and cultural ecologists. Humans impact the environment by utilizing natural resources and by placing material objects on the landscape, including farms, cities, and buildings. This alters the physical appearance of the natural landscape and affects natural processes. The cultural landscape is thus formed and becomes the tangible chronicle of a cultural group and its agricultural traditions, values and belief systems, settlement patterns, social structures, and political practices (Groth and Bressi 1997; Lewis 1979).

Landscapes can be viewed as the alterations to the natural environment and physical structures that result from cultural attachments to place. They are, in Conzen's (1990: 1) words, 'the embodiment of the cumulative evidence of human adjustment to life on Earth.' Landscapes are studied as a phenomenon to be explained, but more importantly, as a means of understanding the societies that have produced them, for scholars acknowledge that 'nature, symbolism, and design are not static elements of the human record but change with historical experience' (Conzen 1990: 2). They are 'historically dynamic places in perpetual evolution, shaped by social values, attitudes and ideologies as they contract and expand, deteriorate and improve over time and space' (Ringer 1998: 7).

Of particular interest in this volume are those parts of the cultural landscape that are created by political systems and tourism. According to Kliot and Mansfeld (1997: 498), the political landscape is one characterized by 'a peculiar geographical association of political facts or recognizable political constitutions' that reflect the political values, ideologies, and legal systems of a place. Elements of the political landscape include public institutions, symbols, and monuments, as well as political territorial behavior. This includes political boundaries and the physical landscapes they create or that are created for them (Jackson 1984). Human behavior, settlement patterns, power structures, and economic activities that are influenced, and even determined spatially by the existence of a political boundary can be considered part of the political, or border, landscape.

Since they separate different political and social realms, boundaries create noticeable differences in the spatial development of various phenomena on both sides. For example, in a prairie region otherwise undifferentiated in natural terms, contrasting cultivating practices, agricultural specializations, and government policies on opposite sides of the US–Canada border have created distinct patterns of agricultural development on either side (Reitsma 1971). In fact, this international border creates 'sharp differences in the areal distribution of crop and livestock production' (Reitsma 1972: 10). Likewise, other studies confirm that the border has a notable impact on patterns of television viewing across political divides (Minghi 1963b), and distinct production systems and patterns of industrial development are also juxtaposed where nations meet (Gebhardt 1987).

Another important point here is that as tourism develops in a destination, a distinct cultural landscape is created – a tourist landscape. Ringer (1998: 6) suggests that, while difficult to define, tourism landscapes are 'the manner in which the visible structure of a place expresses the emotional attachments held by both its residents and visitors, as well as the means by which it is imagined, produced, contested and enforced.' Tourism landscapes then are those created by and for tourists. They are the visible structures that result from tourism's attachment to place, as well as the images, or myths, of place that are produced, contested, and enforced by various agents (e.g. residents, promoters, governments). Several authors have studied the spatial patterns and landscapes of tourism (e.g. Hudman 1978; Oppermann 1998; Pearce 1998; Stansfield and Rickert 1970; Timothy and Wall 1995) and found that they are a prominent form of cultural landscape in destinations, particularly in urban areas. In many cities, tourists and hosts are often concentrated in areas commonly known as tourist districts, where the infrastructure, services, and other physical characteristics have developed largely as a result of tourism (Pearce 1998; 1999b; Stansfield and Rickert 1970; Timothy and Wall 1995).

This chapter uses these concepts to examine border and tourism landscapes. First, since borders and the socio-political systems they enfold create contrasts in spatial and administrative patterns on opposite sides, differences in tourism patterns and landscapes across boundaries are discussed. Second, borders also determine the nature of the tourist landscape and its elements that will develop in frontier regions. Finally, while borders have a stronger effect on tourism, tourism can, and does in many instances, help to create and alter the border landscape.

Contrasting development on two sides of a border

Like the earlier discussion about differences in spatial patterns and land use on opposite sides of a border, researchers also acknowledge that patterns and flows of tourism, as well as the tourism-related landscape itself, demonstrate significant visible differences on opposite sides of a political boundary (Arnould and Perrin 1993; Varniere-Simon 1991). For example, in some places, tourism is heavily developed spatially adjacent to a border on one side but not on the other. Some communities along the US–Mexico and US–Canada borders fall into this category, and examples like the Dominican Republic and Haiti, which share the island of Hispaniola, are vivid reminders of how tourism thrives on one side (Dominican Republic) but not on the other.

Cultures that act as attractions also differ on opposite sides of a border, as do the types of tourism, or tourist activities, that are planned and developed. A clear example of these is the island of St Martin, which is bisected by an international boundary (Dutch and French) (Figure 4.1). While the island is small (approximately 95 square kilometers), and is the smallest island territory divided between two sovereign states, the two sides are culturally distinct. This is a major attraction for tourists, to be able to visit one island with two cultures, and the theme has become a primary selling point in promotional literature. Guide books and travel writers promise an international experience that is 'delightfully Dutch' and 'fantastically French' (Langford 1998; O'Neil 1996; Schrambling 1991; *Travel Weekly* 1991). According to some observers, St Martin is notably French in terms of culture, food, and architecture, while the Dutch side (St Maarten) is clearly Dutch (Chase 1996; O'Neil 1996; *Successful Meetings* 1992; *Travel Weekly* 1991). St Martin is less developed than St Maarten physically in terms of tourism. According to Schrambling (1991: 101), this is one of the island's primary appeals because 'to a certain extent, all those travelers get double their money's worth: two islands in one.' Not only do the cultural landscapes differ between the two sides, the types of tourist activities are also different. The government on the Dutch side has allowed the spread of casinos, but on French soil they are not permitted (Chase 1996; Kersell 1991; Langford 1998). Nude beaches are more common on the French side, but shopping facilities are better developed in St Maarten, and the island's strip bars are located on the Dutch side as well (Chase 1996).

Contrasts in the tourist landscape are not at all uncommon, such as the casinos on St Maarten discussed above. McGreevy (1988; 1991) has written extensively about such differences between Niagara Falls, USA, and Niagara Falls, Canada. His research on the history of the Falls revealed that as early as the nineteenth century, visitors at Niagara were 'struck by the contrast between the landscape of the Canadian side and that of the American side. They found the American landscape messy and tasteless, while the Canadian side seemed vibrant, orderly, almost manicured' (McGreevy 1991: 1).

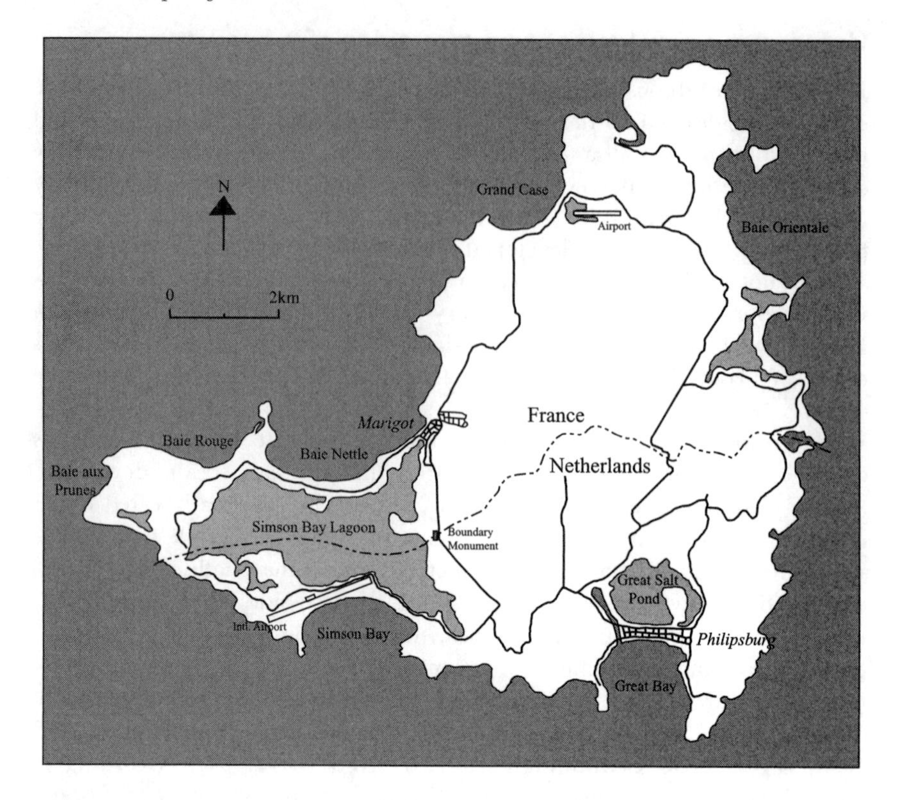

Figure 4.1 Sint Maarten/Saint Martin

Modern-day visitors still note the differences. In 1985, one woman who had visited Niagara Falls wrote a letter to Dear Abby, a syndicated newspaper column, to comment on her experience. 'The Falls on the American side were grossly neglected and looked terrible, but the Falls on the Canadian side were beautiful, bright with flowers and well-maintained' (van Buren 1985, quoted in McGreevy 1988: 308–9). More than 1,900 people responded with letters agreeing with the original writer; fewer than 200 wrote to defend the American side (McGreevy 1988: 309).

These spatial differences in tourism development are noteworthy. The processes and causal relationships that create these contrasts are manifold and complex. Communities separated by international boundaries, regardless of how close they might be physically to one another, experience different histories. This shapes their social values, political institutions, economic conditions, and cultural traditions. While these factors are complex, an attempt is made here to discuss them under two headings: cultural and socio-economic differences, and differences in planning traditions and policies.

Cultural and socio-economic differences

In areas where cultures and societies are different on opposite sides of a border, the nature of tourism will also probably be different to some extent. Religious differences and social traditions, for example, may determine the types of tourism that develop on opposite sides, or at least the nature of the tourist activities that are permitted to grow. Settlement and land tenure patterns, as well as socio-economic differences also affect how tourism grows on opposite sides.

Settlement patterns and urban structures

Settlement patterns, land tenure systems, and urban structure commonly vary from culture to culture and across political divides. For example, a glance at an aerial photograph or satellite image of the Czech Republic–Austria border region reveals significant land use differences. On the Czech side, agricultural plots are large and symmetrical, owing largely to the collectivized farming practices established by communist rulers. On the Austrian side land plots are smaller and asymmetrical.

Cities divided by borders tend to evolve as two separate urban settlements (Herzog 1991a). This usually results in differing urban structures on each side. Urban development on one side might be crowded and unplanned, while on the other, orderly and well planned. Cultural traditions can also significantly affect the layout of cities. For example, Latin American cities have traditionally included centrally located open spaces in the form of squares, or *plazas*, which have considerable cultural meanings. A linear spine of commercial activities offering high-order goods and services is also common in Latin American city structure, and has cultural and economic implications (Griffin and Ford 1980). US cities are not as culturally bound in their physical composition. This creates a unique contrast in transfrontier urban areas along the US–Mexico border, particularly in relation to tourism. As Curtis (1993: 64) put it, 'though physically close and functionally interrelated, distinct economic, political, and social processes and consumer behaviors give rise to contrasting urban forms and patterns of spatial organization.'

Socio-economic differences

Economic disparities on opposite sides of an international boundary will determine the rate of tourism growth and the efforts expended by governments and local residents to develop the industry (Klemenčič and Gosar 1987). If Country A has more money to spend on promotional efforts and infrastructure than Country B, distinct patterns will likely arise between the two countries' tourism industries.

By way of example, the French part of the Ardennes Mountains experienced rapid industrial development in the nineteenth century, while the Belgian part of the region remained rural, traditional and underdeveloped into the twentieth century. Tourism is being targeted in both regions, but the French are focusing on urban and industrial heritage, while the Belgians focus more on rural tourism (Varniere-Simon 1991).

In the context of Cyprus, the South has managed to build a highly developed economy within a relatively short period of time. While less developed, the North has improved but is still completely reliant on the mainland Turkish government. International support for the South helped that part of the island recuperate, while sanctions against the North have prevented widespread and rapid development (Kliot and Mansfeld 1997: 516). International economic aid poured into the South after the events of 1974, enabling a rapid economic recovery. The North, on the other hand, has been reluctant to use tourism as a major catalyst for economic development owing to the international boycott and its overdependence on the fragile Turkish market (Mansfeld and Kliot 1996).

Tourism developed much more slowly on the French side of St Martin (Chardon 1995; de Albuquerque and McElroy 1995) than on the Dutch side. The Dutch began developing their tourist services early on when they targeted the mass market of North American and European middle and upper middle classes. They built the airport and cruise ship port, and they licensed casinos and promoted duty-free shopping first (Kersell *et al.* 1993: 61). Dutch marketing efforts were quick, but the French did not try to compete by duplicating the services. Instead they focused on complementary services such as high-class restaurants, boutiques, and villages, which appeal to the more discriminating visitors than the masses that the Dutch authorities were attempting to lure. The French side also catered more to the small chartered boats and private yachts rather than large cruise ships (Kersell *et al.* 1993: 61).

Planning traditions and policies

Simply stated, Chardon (1995) suggests that differences in tourism across borders are directly linked to space being separated by two different sets of laws. This appears to be the case, and legal differences manifest themselves in tourism policies and practice.

Opposing ideological/political systems

In regions where diverging political systems come together, tourism will be influenced by differing ideologies. A clear example is Finland and the Soviet Union, where in the former tourism flourished because people were permitted to travel freely, while in the latter, most tourists were permitted entry only in groups, and their routes and itineraries were predetermined and strictly adhered to. Similarly, until their unification in 1990, the boundary between North and South Yemen separated two very different philosophies and sets of priorities towards tourism. The 'anarchic' capitalist North was more interested in developing and promoting tourism than was the communist South, with the effect that North Yemen had more experience and expertise in planning and managing the industry (Burns and Cooper 1997). Although the two countries were unified in 1990, many of these differences still persist, as administrators and management approaches remain based on traditions. As a result, the future over-exploitation of tourist resources in the North are likely, while the South will probably remain a marginal tourist destination (Burns and Cooper 1997: 562).

In terms of economic development in West Germany, the government provided aid programs to people in the border zone. Also, large sums of money were spent on reconstructing and modernizing villages and towns in terms of infrastructure, heavy industry, and tourism. In the East, however, the border zone declined economically, and the government heavily restricted the flow of people within the buffer zone, and did little to encourage economic and infrastructure development in the area for fear of contamination from, or escape to, the West (Buchholz 1994: 58–9). These conditions created significant challenges to the Federal Republic when the two countries were united in 1990. Roads, railroads, water pipes, sewerage systems, and electricity supplies had to be completely rebuilt (Buchholz 1994).

Opposing political systems and ideologies on the island of Cyprus have created other landscape differences. One of the first things to be done by the Turkish rulers of North Cyprus was to rename all of the place and street names that had previously been Greek and destroy Orthodox churches or transform them into mosques or museums. In the South, however, Turkish street and place names are preserved, mosques are well maintained, and a special committee oversees Turkish property, 'keeping it intact for the day when its owners will return to claim it' (Kliot and Mansfeld 1997: 511).

Zoning and land use policies

Zoning regulations are influenced by international boundaries. Urban zoning regulations in one city may create an environment of conservation and urban renewal, while across the border cities might be facing rapid deterioration with little regard for conservation efforts. Regulations regarding building size and intensity of development also vary from one political jurisdiction to another, thereby creating distinct urban and tourism landscapes (Herzog 1991a). In the same way, differing rural land use and natural resource conservation rules create distinct visual differences in these landscapes on opposite sides of an international divide.

For instance, Turkish Northern Cyprus has fewer planning regulations than the South as regards urban and resort development. Development laws in the Greek South aim to preserve the scenic views and natural landscape, which has resulted in beachfront development being more controlled there than in the North. Northern developers are permitted to build hotels and other tourism infrastructure as close as 50 meters from the shoreline (Mansfeld and Kliot 1996: 198).

Likewise, differences in urban planning ordinances along the US–Mexico frontier, together with the cultural differences alluded to earlier, have created a distinct transfrontier spatial pattern. According to Curtis (1993), Loredo, USA, and its neighbor Nuevo Loredo, Mexico,

> are very different places despite their proximity and history of interaction. The differences show up readily in visual elements such as architectural styles and the use of exterior color and embellishments; in the overall condition of the buildings; in the maintenance of the street and sidewalks; in the landscaping of the public, commercial, and residential spaces; and in the intensity of land utilization.
>
> (Curtis 1993: 56)

Regional planning traditions

Planning structures on opposite sides of a boundary can affect the way tourism functions and the way it develops on the ground (Arnould and Perrin 1993). In some situations, planning on one side of a border may take on a more top-down, centralized approach, while in neighboring jurisdictions it is done at a more grassroots level with greater input from citizens and lower-order polities. Participatory, bottom-up planning tends to be less common in the developing world, where national governments play a more central role in development processes. This contrast is particularly striking in situations where traditional societies border modernized and developed nations. For instance, planning in Mexico is a much more centralized function, highly controlled at the national level (Kjos 1986), whereas in the United States, development operations tend to be planned and administered at more local levels (i.e. state and county), and public participation is more common.

Similarly, while cooperative tourism planning is a rather new concept in most parts of the world (Timothy 1998b), it is clear that if adjacent jurisdictions will collaborate on issues of economic development and conservation, such as tourism, the industry itself and the local environment will have a better chance of being sustainable. This will be discussed in greater detail in chapter 6. In common with many destinations, there is a lack of coordinated planning between officials of St Martin and St Maarten, which results in very different approaches to tourism development. The French are inclined by political traditions to be rational and to plan public-sector activities. The Dutch inclination, however, is towards a *laissez-faire* attitude in both government and business (Kersell *et al.* 1993: 51). This has resulted in French officials giving up on trying to involve their Dutch counterparts in most development projects, and a tendency for tourism in St Maarten to be guided more by unplanned and uncontrolled private initiatives (Kersell *et al.* 1993). This likely explains many of the differences between the two parts of the island in terms of types of tourism and tourist activities as well as cultural differences and levels of development.

The political perspectives best explain the differences in tourism development on the two sides of Niagara Falls. From the beginning, the Canadian government controlled a narrow strip of land along its side of the river for military purposes, which, according to McGreevy (1991: 3), made it simple to implement an organized landscape design when the notion of public parks emerged in the 1880s. In Canada, the government has also long dominated hydroelectric production and, as a result, the government power commission has been able throughout history to coordinate its development projects with the Niagara Parks Commission. On the US side, however, private enterprise has led industrial development, so that coordination and preservation have been more difficult (McGreevy 1991).

Furthermore, heavy industry began sooner and grew more quickly on the American side. Canada also possessed significant industrial development, but there it was located further from the border. Rather than heavy industry, Canadian Niagara focused its efforts more on tourism, which by 1880 had become the primary industry on that side of the border. According to Donaldson (1979: 117 quoted in Getz 1992), 'In the 1820s Canadians were grabbing the burgeoning tourist trade while Americans lined their river bank with factories.' When it was realized that uncontrolled commercialization had grown to disgraceful proportions, the New York city created a small state park near the Falls in 1885. Soon after, the Ontario side took action because it was discovered that Canadians favored the American side over their own (Getz 1992: 757). As a result, the Niagara Parks Commission was created in Ontario later in 1885 (then called the Queen Victoria Niagara Falls Park Commission), with 62.2 hectares of public parkland adjacent to the Falls. Gradually more and more land came under provincial protection until now it comprises over 1,250 hectares near the Falls and along the river. Since 1887, the Commission has been a major conservation and planning force on the Canadian side, which has resulted in the public opinion that the Canadian Falls are more attractive and scenic (Getz 1992; 1993). The US parklands were more limited spatially, so that by the end of the nineteenth century, heavy industry on the American side had developed so rapidly that dozens of mills had accumulated along the gorge, nearly to the brink of the Falls (McGreevy 1991: 2; Schwartz 1997). By the Second World War, Niagara Falls, New York, had become a major electro-chemical center, and the city's air had turned brown with grit and toxic fumes (McGreevy 1988).

McGreevy (1991: 3) argues that part of the difference between the two sides of Niagara Falls is that the place had a different meaning for each country. If state and federal officials in New York had decided that Niagara Falls was a valuable national treasure, as was the feeling in Canada, they could have worked to create a more pleasing landscape. The difference, he suggests, is that for Canada, Niagara is

> like a front entrance. It has been embellished and manicured. To find the part of the lot that corresponds to America's Niagara, we must travel around the house to the back alley. There, behind the garage, is a row of garbage cans. This is a utilitarian place to which we are not meant to direct our attention. Niagara is like a dark, back alley of the United States, an appropriate place for dioxin and nuclear waste, but apparently not for gardens.
>
> (McGreevy 1991: 4)

While tourism services on the Canadian side were dispersed and the zones immediately adjacent to the Falls were managed as parklands, on the US side, formal planning did not occur until the early 1970s, prior to which the city was faced with old tourism infrastructure, a bad reputation, and a declining industrial base (Getz 1993). Recent planning efforts on the US side have created a heavy concentration of clustered tourism businesses mingled with resident services, or CBD functions, creating what Getz calls a tourism business district (Figure 4.2). On the Ontario side, the Niagara Falls landscape still resembles more of a recreation business district, where clusters or linear aggregations of restaurants, food services, stores, and souvenir shops dominate (Getz 1993; Stansfield and Rickert 1970).

Borders and the spatial development of tourism

Border crossing points are magnets for the physical development of tourism (Cosaert 1994; Klemenčič and Gosar 1987). Fernandez (1977: 123) argued that since tourist dollars are, directly or indirectly, the main source of the personal income of most residents of many border towns, in his case Mexico, it is tourism that 'gives to the gaudy commercial enterprise the glaring predominance over other aspects of economic life and is, therefore, the most immediately perceivable element of economic life in the border towns of Mexico.' Transit traffic (discussed below) creates a unique physical landscape as people pass through border towns. Similarly, when the border is the primary attraction the crossing points become focal points of spatial development, although as Lösch (1954) and Mackay (1958) pointed out, the boundary effects do diminish with distance from the border. As such, the

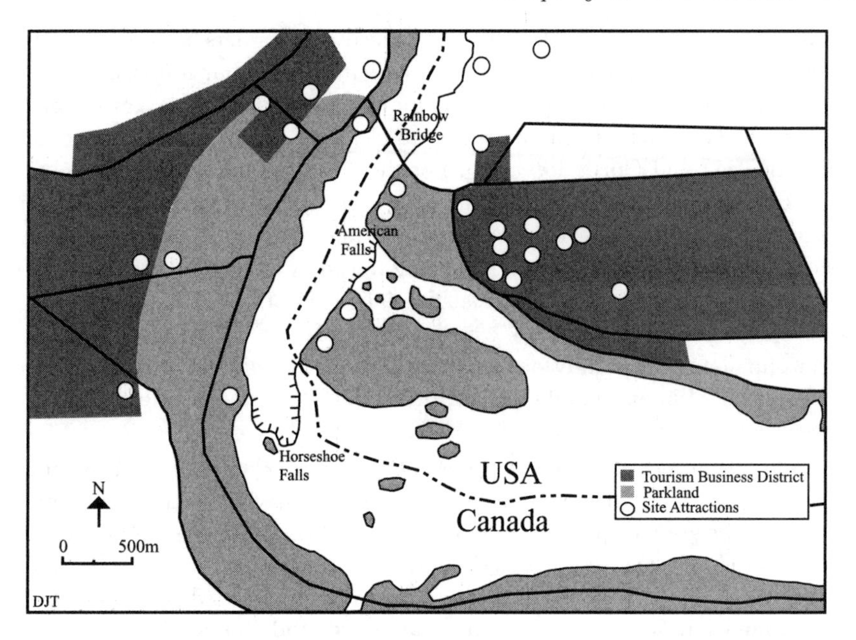

Figure 4.2 Tourism business districts – Niagara Falls
Source: after Getz 1993. Reprinted with permission of Elsevier Science

general trend is that the closer to the border, the more intense the development will be for both transit and destination traffic. Impermeable boundaries also create landscapes where services and infrastructure run parallel to each other and they produce distinctive patterns of traffic flow.

Borders as destinations

The tourist landscape on the US–Mexico border is shaped by a primarily pedestrian traffic flow, while on the US–Canada border, the spatial arrangement is shaped more by vehicular traffic, so that tourism services are located slightly further inland from the crossing points. Both of these examples are described in this section.

Curtis and Arreola's (1989; Arreola and Curtis 1993) extensive research analyzes the development of tourist districts along the US–Mexico border. Much of the information in this section comes from their accounts. In frontier cities, tourist districts are usually found along the streets adjacent to, or within easy walking distance of, the border crossings. According to Arreola and Curtis (1993: 82) in the Mexican context, this clearly reflects the popular belief among Americans that border towns are dangerous places to visit and drive. Some people are intimidated by unfamiliar traffic laws and signs, or they fear the consequences of an accident or having their vehicles damaged or stolen. Legends also exist of police who accuse foreigners unjustifiably of mistakes they have not made in order to elicit bribes from tourists. Others may wish to avoid spending time and money getting Mexican automobile insurance or long delays in line waiting to clear US customs on their return home (Arreola and Curtis 1993: 82). For these reasons many tourists choose to park their cars in the United States and cross the border by foot. The locations of tourist districts in Mexico near border crossings are convenient for visitors who feel more comfortable knowing that they are within easy reach of the United States (Arreola and Curtis 1993; Timothy 1995a). Since the Prohibition era (1918–33) most US visitors to Mexican border towns have been excursionists, generally staying from four to eight hours in their destination in search of booze, gambling, sex, other exotic diversions, and shopping.

> In all likelihood, few tourists come to these [border] cities for sight-seeing purposes, although 'negative sight-seeing,' or viewing disturbing scenes such as begging or poverty-ridden living conditions, may be a motivation for some. Rather, we contend, most tourists want to eat, drink, and purchase selective goods and services in an idealized 'Mexicoland' that is somewhat insulated from the real world and danger but one that evokes at least a veiled sense of excitement and foreign adventure.
>
> (Arreola and Curtis 1993: 92)

These characteristics, together with tourist anxieties about visiting the border towns, have contributed to the creation of tourist districts that are conveniently located near border crossings, small in size, and somewhat isolated from other parts of the city (Curtis and Arreola 1989: 24) (Figure 4.3). The tourist districts are overwhelmingly pedestrian-oriented, which forces visitors to interact more personally with locals and the environment around them. For some tourists this may be part of the appeal, while for others it may be disturbing (Curtis and Arreola 1989: 25). In most Mexican border towns, the tourist district has historically been the focal point of urban development. This has been the case in Tijuana since its inception (Herzog 1985; 1991b).

Generally, the goods and services that appeal to most visitors, such as curio shops, clothing stores, liquor stores, bars, and restaurants, are located near the border crossing point and dominate the main tourist strips (Herzog 1985). Medical/dental offices, which are also important for American visitors, tend to be located further from the border crossing, as are shops catering mostly to Mexicans, such as furniture, appliance, and drug stores (Curtis and Arreola 1989). Nuevo Loredo (Mexico) has significantly more medical offices (e.g. physicians, opticians, dentists) than does Loredo (USA), many of which cater to a US clientele because of the low cost, as witnessed by window signs stating 'English spoken here' (Curtis 1993: 61). For Mexican patients, the location of these facilities is related to accessibility, particularly with public transportation.

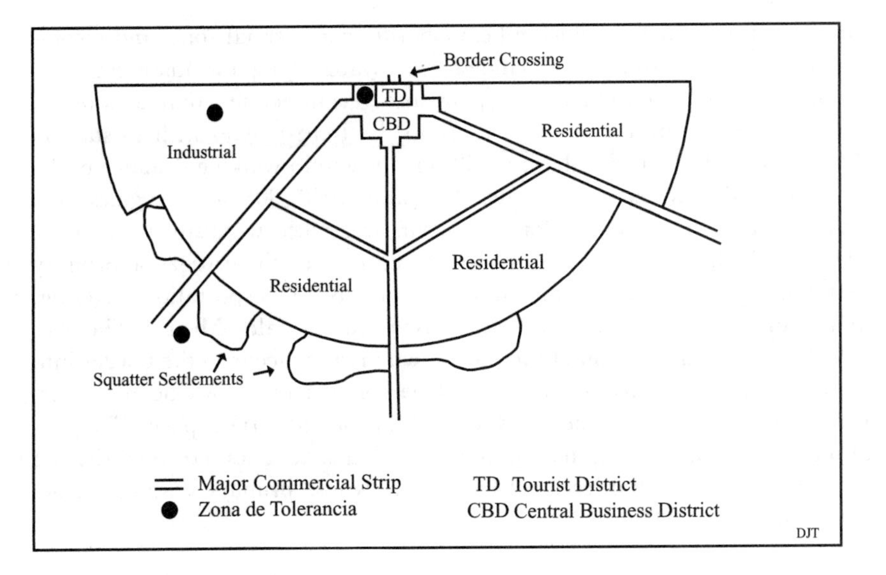

Figure 4.3 Tourist districts in frontier urban areas
Source: after Arreola and Curtis 1993. Reproduced with permission of Arizona Press

Another component of the tourist landscape is street vendors, which are particularly ubiquitous in some border tourist towns (Pertman 1995). In Mexican border towns, street vendors operate out of makeshift stalls or spread their good out on blankets. Other ambulatory vendors sell candy, gum, souvenirs, artificial flowers, and other trinkets (Arreola and Curtis 1993: 87; Arreola and Madsen 1999).

While tourist zones are less distinct on the US side, they are still discernable. The primary difference is the dominance of electronic appliance stores, clothing and shoe stores, and jewelry shops immediately adjacent to the crossing point (Curtis 1993), with an absence of restaurants and medical facilities. Eating and medical services are not popular among Mexican shoppers because these are cheaper in Mexico. As discussed in chapter 3, Mexicans visiting US border towns tend to be more serious shoppers in search of bargains, while Americans visiting Mexican border towns are generally more interested in souvenirs. Most retail establishments on the US side are more concentrated nearest the international crossings than in Mexico, which reflects the restricted spatial mobility of the customers and the nature of the goods they purchase and have to carry home to Mexico (Curtis 1993). This creates a stronger distance-decay effect on the US side (Herzog 1990).

Nogales, Arizona, for example, contains a long strip of shops next to the border crossing (Arreola and Madsen 1999; Smith and Malkin 1997), which coincides with the observations of one early urban planning document for the city (Gonzales Associates 1970). The plan described the current infrastructural situation, which included a heavy border commercial zone, and recommended that this zone be expanded northward along the main highway leading from the boundary. The plan also recommended that a hotel be built next to the frontier line to 'cater to the tourists who wish to stay on the Arizona side of the border but still remain within walking distance of the activities within Mexico' (Gonzales Associates 1970: 30). The Nogales plan was unique in that it focused on one urban area with two distinct functions, each of which required its own infrustructure: (1) to serve as a primary destination for Mexican visitors and (2) to act as a transportation, catering, and accommodation center for Americans visiting Nogales, Mexico. The plan called for a transformation of the downtown area adjacent to the border into a structured tourist district that would be convenient for Mexican visitors, providing numerous commercial sites and lots of pedestrian space. The plan also called for the development of more parking spaces, less crowded driving conditions, and the border hotel for visitors whose primary destination lies across the border (Figure 4.4).

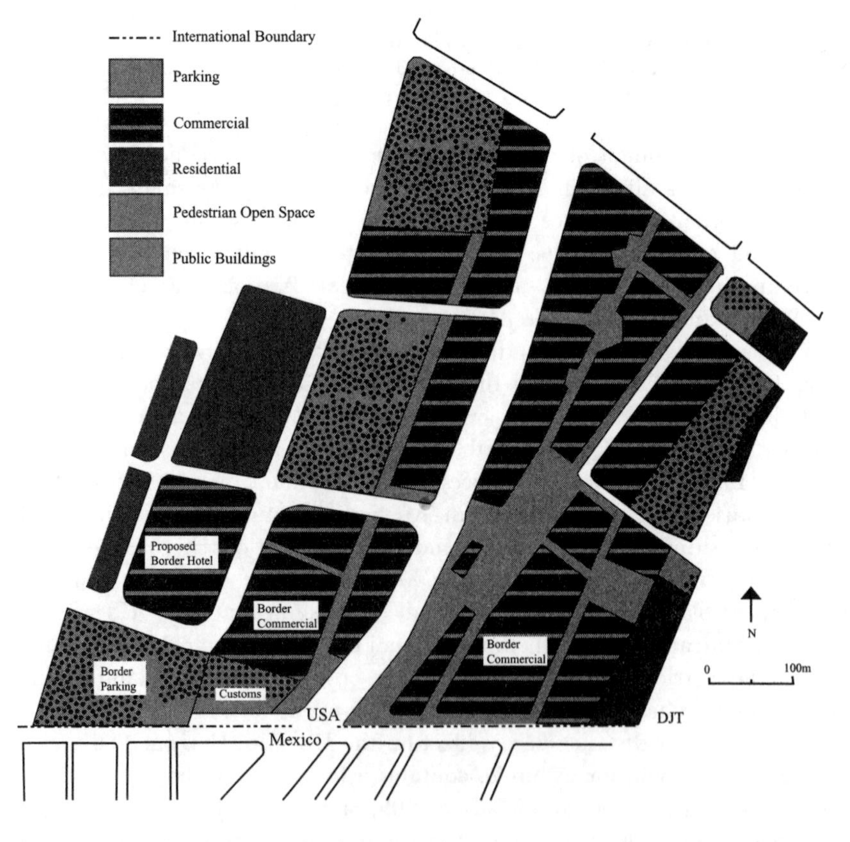

Figure 4.4 Nogales, Arizona, development plan
Source: after Gonzales Associates 1970

While these tourist zones themselves are landscapes of tourism created by boundaries, they also encompass micro-landscapes within them that are characteristic of border regions. In the Mexican communities, symbols such as ornate signs, tile mosaics, and pictures provide important clues about the meanings of tourist districts (Arreola and Curtis 1993: 90). The most common images are *serapes*, *sombreros*, bullfights, mission-style churches, and desert landscapes with cacti. Men are portrayed as sleeping *campesinos* with their knees in front of them and their *sombreros* pulled down over their faces. Women are typically portrayed as high ladies-of-Spain wearing frilly, low-cut, off-shoulder dresses (Arreola and Curtis 1993: 90). These romantic, stereotypical images visually

> symbolize a distinctive American way of thinking about Mexico, and as such may reflect the deep-seated cultural and historical forces that have shaped the landscapes associated with these districts as well as shed light on the attraction they hold for the average North American tourist.
>
> (Curtis and Arreola 1989: 27)

> Indeed, the built environment of these districts as well as the goods and services offered there, play on the preconceived images of what most North American tourists think Mexico *ought* to look like and what it *ought* to offer them the opportunity to buy.
>
> (Curtis and Arreola 1989: 20)

Another type of tourist zone can also be found along the Mexican border, either as part of the regular district or more distant from border crossing points. Curtis and Arreola (1991) found that 12 of the 18 Mexican border towns they surveyed possessed *zonas de tolerancia* – designated zones in urban areas where prostitution is spatially restricted. Two types of *zonas* were identified along the border: district and compound. District *zonas* are areas in a city with a clustering of commercial establishments, such as bars, nightclubs, restaurants, and hotels, where prostitution is tolerated by local authorities (Curtis and Arreola 1991: 335). This type of *zonas* is generally located in the old city core on the margins of the traditional tourist zones. The sizes of the districts range from 10 to 15 blocks in the large cities and down to one block in some of the smaller communities. In some cities, efforts have been successful to relocate *zonas* to less conspicuous locations (Arreola and Curtis 1993; Curtis and Arreola 1991). The second type of *zonas*, compounds, differ from the district type in terms of location, landscape characteristics, and organization. Typically these have developed away from the city center in areas of relative isolation. Often compounds are located on the outskirts of town, and they are characterized by walled areas, a single entry gate, armed police protection, clubs, bedrooms, restaurants, pool halls, liquor stores, tattooing services, and food and drink stands (Arreola and Curtis 1993; Curtis and Arreola 1991).

In common with the Mexico–US example, intense tourism services developed along the French–Belgian border, usually clinging to crossing points. Shops, cafés, gas stations, and restaurants developed adjacent to border crossings, creating on the Belgian side commercial strips in urban areas and clusters of services in rural areas that catered primarily to French consumers (Cosaert 1994).

Since Canada–US border traffic is characterized more by vehicular travel, the landscape there differs slightly from the situations described above. Tourist facilities (e.g. shopping centers) tend to be located further from the border crossing, reflecting higher levels of mobility. The landscape of cross-border shopping is unique and is completely at the mercy of fluctuations in exchange rates and other economic variables. When rates are favorable, shopping facilities are built rapidly within easy access of the border, and when rates are unfavorable, these same facilities close down.

In the heyday of Canadian shopping in the United States, additional shopping malls, grocery stores, and hotels were built in American towns all along the border, and many existing businesses were expanded (Scanian 1991b; Stevenson 1991). Stores, hotels, cinemas, gas stations, museums, and banks all benefited from this consumer craze (Rinehart 1992). However, with the decline of Canadian shopping in the US, the scene has changed. Observations by the author in July 1998 found that dozens of grocery stores, shoe outlets, restaurants, gas stations, and department stores in many American communities along the border have closed down. The shopping landscapes of the early 1990s have become derelict reminders of the short-lived zenith of Canadian consumerism, and communities that became too dependent on dollars from their northern neighbors have become virtual ghost towns. Bradbury and Turbeville (1997) note similar trends in the western United States. Conversely, on the Canadian side, shopping facilities are being expanded and built, albeit with more caution than that exercised by the American neighbors.

Visual clues abound in border communities that testify of the importance of cross-border consumerism on their local economies. For example, Canadian and US flags flying over shopping malls, stores and restaurants advertising the current exchange rate, and even signs and billboards openly welcoming Canadians dominated the landscape of cross-border shopping in US border towns in the early 1990s (Bradbury and Turbeville 1997).

As mentioned in chapter 3, many US businesses began accepting Canadian currency, and visitors had the option of paying in either currency. This resulted in the two-drawer cash register in many US border communities: one for Canadian dollars and one for American dollars (Bradbury and Turbeville 1997: 14). Turbeville and Bradbury (1997) postulated that there are distinct patterns of morphology in retail, entertainment, and commercial sectors of the US and Canadian border towns which are deeply rooted in their respective national consciousness. American border towns are much more replete with garish displays of binationalism than their Canadian counterparts, while the Canadian communities demonstrate a far greater influence of the border on local image ('Canadianness') and landscape, and it is clear that they attempt to ignore the border altogether.

The same type of landscape is created in towns adjacent to national park entrances. Since infrastructure development is severely curtailed in national parks, such services and structures develop just outside park boundaries. Towns have developed rapidly along approaches to national parks and usually provide the last chance to fuel the car, shop for groceries, eat a restaurant meal, or find a place to sleep. These towns have become known as gateway communities, and their primary function is to fulfill the needs of tourists – needs that cannot be readily fulfilled within the park (Steffens 1993). Unfortunately many gateway communities have been overdeveloped to the point that they begin to diminish the ambience of the park itself. Steffens (1993) describes many of the problems facing park managers as intense development outside park boundaries encroaches upon the parks, thereby influencing the panorama and the wildlife.

Borders as transit zones

Perhaps the most common relationship, but the least understood, between borders and tourism is that of boundaries as lines of transit. Millions of people cross boundaries every year for a variety of reasons, some of which keep them in the frontier zone, such as shopping and the vices described earlier. Other people travel through and past borders on their way to more distant locations. The discussion in chapter 3 shows how borders and their surrounding regions can function as tourist destinations. However, from a broader view of global tourism, relatively few people ever regard the border as more than a line of transit, a place to go beyond, and a nuisance (Battisti 1979; Gosar and Klemenčič 1994; Matznetter 1979). From this perspective, borders exhibit the least significant visible impacts on tourism. Once through crossing procedures or past the welcome sign, most travelers hastily continue on to their various destinations. Unless some attraction exists near a port of entry to capture the interest of tourists, few, if any economic impacts from transborder travel will be found in border communities (Timothy 1995a: 149).

There are exceptions to this general statement, however. Owing to advantageous location, several types of services commonly locate near ports of entry. Duty-free shopping is one of the most important elements of the transit landscape associated with border crossings (Plate 4.1). Nearly all international airports and harbors provide tax-free services to travelers once they have cleared departure inspections. Similarly, duty-free shops at land crossings are usually located between the customs depots of each country. By buying goods after they leave the effective control of one country but before entering the next country travelers can avoid paying taxes and import duties

Plate 4.1 The marker on the left marks the actual location of the Canada–USA border. This duty-free shop is located only five meters inside Canada

on many highly taxed items such as liquor, perfume, leather, jewelry, and cosmetics (Lyons 1991).

An additional feature of the transit landscape is welcome centers, or tourist information offices. Most states and provinces in North America have established welcome centers along primary highways at or near their boundaries, and many countries in Western Europe have done the same. In addition to providing travelers with a convenient place to rest and relax, these facilities promote tourism within the state or province by disseminating information about attractions, restaurants, and accommodations. By placing this type of establishment near interstate boundaries, a tourism organization can encourage travelers, before going any further, to see more of the state and to utilize its resources better (Howard and Gitelson 1989). Several studies have found that use of restroom facilities is the primary reason for stopping at state welcome centers, followed closely by obtaining tourist information (Fesenmaier *et al.* 1993; Howard and Gitelson 1989; Muha 1977). In terms of information, research has shown that the use of welcome centers has significant impacts on travelers entering a state or province. In their Indiana study, Fesenmaier and Vogt (1993) found that 21 percent of the people they surveyed stayed in the state longer than originally planned, 29 percent visited places not planned prior to stopping at the welcome center, and 33 percent indicated that they spent more money than anticipated as a result of their stop at the highway information center. Tierney (1993) similarly concluded that people stopping at Colorado centers resulted in a 2.2-day average increase in length of stay and over US$1 million in additional spending because nearly half of all travelers' decision to visit an attraction were made after they entered the state. Most of the people who use welcome centers originate in non-adjacent states (Stewart *et al.* 1993), probably because they are unfamiliar with the state they are entering and desire to learn more about it and what it has to offer. Gitelson and Perdue (1987) concluded that, in addition to using the information obtained on the current trip, many travelers rely heavily upon the literature acquired for planning future trips as well. A later study drew the same conclusions (Fesenmaier *et al.* 1993). It is clear that strategically placed information centers at state and provincial boundaries along highways which draw substantial numbers of tourists, can enhance the tourist experience, and can be effective in increasing length of stay and levels of spending by out-of-state travelers (Timothy 1995b: 530).

Other types of services tend to locate around border crossing points, including gas stations, restaurants, currency exchange booths, and accommodations (Bachvarov 1979; Smith and Malkin 1997), even when the communities are not significant tourist destinations. Crossing a border may be viewed as a milestone in the progress of a trip. Therefore, these businesses provide rest stops for people entering a territory who might be in need of their services and cluster near crossing points to take advantage of the flow of incoming traffic. While transit border towns can benefit from high levels of tourist traffic, Pagnini (1979) suggested that they generally do not generate profits commensurate with the overuse of local infrastructure and other stresses that befall them. These locations, he suggests, ought to receive higher development priority from national governments, since they suffer the consequences of high flows of traffic taking the bulk of its spending further toward the interior of the country.

Similarly, the island nation of Singapore has developed into a gateway for much of Southeast Asia. This concept resembles a spoke-and-hub idea where the gateway, or hub, fans out to numerous destinations within the region. This role as transit gateway has not only created a broad regional spatial pattern of dispersal, it has also created within Singapore a strong service-oriented landscape that provides for the needs of transit tourists who spend several hours, or one or two days, in the country before heading on to their primary destination (Low and Heng 1997). Accommodations, entertainment, recreation facilities, an expanded airport capacity, and well-maintained telecommunications facilities all form part of this gateway milieu.

Parallel tourism development

The existence of a hostile border creates a parallel system of facilities, services, and infrastructure (Kliot and Mansfeld 1994). This is particularly evident in international urban areas where the border is either impermeable or semipermeable. This situation will nearly always result in a lack of cross-border interaction and will deter traffic flows in a variety of ways.

Two examples of borders that do not allow a great deal of cooperation and traffic flow are Cyprus and the former East–West German frontiers. As a result of the Cyprus partition, each entity had to redevelop installations that had been lost to it, and many tourism-related services in the two halves of Nicosia were rebuilt back to back. According to Kliot and Mansfeld (1997: 508), this resulted in a 'dual political landscape of infrastructure and development projects.' For instance, the Republic of Cyprus had to construct a new airport at Larnaca; the North did the same at Erçan (Kliot and Mansfeld 1997; Mansfeld and Kliot 1996). Likewise, the main tourism centers on Cyprus were lost to the Turkish North, which resulted in the South having to redistribute the tourism infrastructure as soon as possible in order to reactivate the industry (Mansfeld and Kliot 1996), thereby creating a dual tourism system on one island completely separated by the border.

In like manner, although Berlin prior to 1990 was one city divided, all institutional links were severed shortly after the Second World War, which resulted in nearly every public institution (and many private ones) being duplicated: opera houses, hotels, symphony orchestras, universities, state libraries, museums, and zoos (Elkins and Hofmeister 1988; Ellger 1992).

As mentioned above, where borders are less restrictive, back-to-back, or parallel, development will be less obvious because cooperative planning may exist and free flows of people are not hindered. Leimgruber (1998) describes four cases of international cooperation along the Swiss frontier. In the example of Basle, he demonstrates how the three countries (Switzerland, France, and Germany) that form the Basle international conurbation, known as the RegioTriRhena, rely on each other for tourism and recreational services, and since tourist traffic is not hindered, little parallel development occurs. Likewise, Timothy (1999a) demonstrates how cross-border cooperation along the US–Canada border results in fewer overlapping services. This results in more efficient management of tourism resources because adjacent regions can rely on the their foreign neighbors to provide some of the services necessary in a regional context. This concept will be discussed at greater length in chapter 6.

Patterns of tourist flows affected by borders

From the earlier examination of real and perceived barriers, it is clear that subjective interpretations of the phenomenal environment have significant implications for human interaction across borders generally and in particular for the spatial development of tourism.

Reynolds and McNulty (1968) also suggested that when societies are not allowed to extend freely across a political divide, personal action space is likely to be skewed in directions along, or away from, the border. This is particularly true when communication channels have been developed inwardly to the exclusion of trans-boundary interaction. Furthermore, border crossing points function as funnels, concentrating the traffic from wide areas on both sides (Figure 4.5), and thus, the spacing of these ports of entry can significantly influence the perceptual environments and spatial behavior of border zone residents and travelers. Individuals who live near a boundary but far from a crossing point are less likely to include areas on the opposite side of the border within their action spaces than people who live near a crossing point because they are functionally farther from the other side (Reynolds and McNulty 1968: 33). As Leimgruber (1989: 51) suggested, even a short distance can be an obstacle to interaction if there is some physical, institutional, or psychological barrier, just as a considerable linear distance can have little effect if no such obstacles exist. For people who live at the border but far from a crossing point, the border might be viewed as impassable. For border residents who live near a crossing point, however, the border is probably more likely to be viewed merely as a nuisance, but something that can be crossed (Clark 1994).

It is clear that these spatial patterns affect regional tourism from a broader pespective. Vuoristo (1981: 241), for example, noted the spatial incongruences in the patterns of east–west tourism prior to 1989. He suggested that 'an essential difference between both blocks is that practically no tourists come to Western Europe from [the] east while there is some noticeable tourist movement from west to east. The ideological boundary is thus like a semipermeable membrane: it allows penetration from the other side, but not from the another (sic).' The existence of the 'Iron Curtain' and its restrictions on international travel created a pattern of north–south traffic as motorways developed between the 1960s and 1990s. Now that the countries of Eastern Europe are more accessible, however, east-to-west transportation linkages are becoming a major focus of European Union development efforts (Burton 1994).

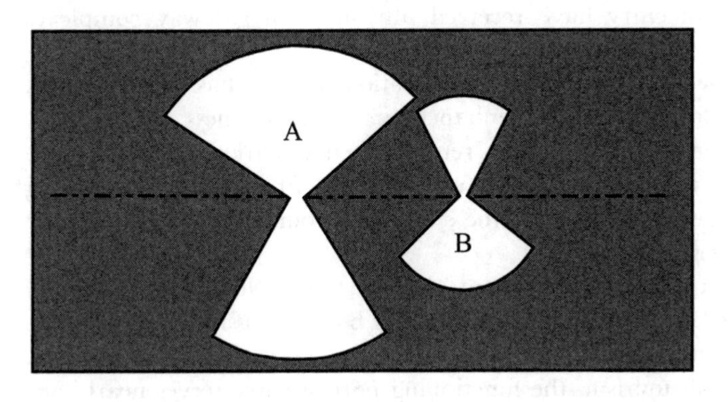

Figure 4.5 The funnel effects of large (A) and small (B) border crossing points

The political landscape of the East–West border itself was remarkable, but the erection of the iron barricade had a significant impact on the tourism infrastructure of the Germanies. The border interrupted ten main railway lines, 24 secondary rail lines, 23 motorways, 140 regional roads, and thousands of smaller roads and paths (Buchholz 1994). Similarly, in the West German city of Hof, the division of Germany changed the function of the town from a centrally located transportation node, with important economic linkages to a large hinterland, into a less important center four kilometers from the border, which severed its economic contacts to the east (Ritter and Hadju 1989: 340).

Tourism's effects on the border landscape

Up to this point this chapter has examined the effects that borders can have on tourism. Most of the time, this is the nature of the relationship, but it is also possible for tourism to influence the nature of the border landscape as well.

Infrastructure

Since the creation of the Programa Nacional Fronerizo (PRONAF) in 1961, much of the Mexican border landscape has been modified and improved. This program included the beautification of the border cities as one of its primary objectives (Dillman 1970a), and massive clean up efforts in the tourist zones has taken place. This was done as a means of cultivating a greener and more wholesome image of border towns and Mexico as a whole, since for many visitors the border towns are their only experience in Mexico. This was an effort by the national government to integrate the border regions better into the national economy, to refurbish urban areas, and to raise local standards of living (Dillman 1970a; 1970b). The plan also called for the development of attractive gateways into Mexico, and today nearly all ports of entry have received ultra-modern gateway complexes (Plate 4.2).

In Tijuana, the main tourist zone, Revolution Avenue, has been widened, repaved, and landscaped to enhance its attractiveness to tourists. Furthermore, storefronts have been renovated, new parking facilities have been built, and a recently built pedestrian walkway from the border crossing to the tourist zone are evidence of the effects that tourism has on the border landscape (Herzog 1985: 301).

As a boost to economic development, the US state of New Mexico recently petitioned the federal government for two new border crossings with Mexico. This resulted in one functioning crossing point and one in the planning stages. Owing to Mexican tourism, the functioning port of entry serves nearly one million northbound vehicles a year, which have a significant impact on New

Plate 4.2 As part of its border development efforts, Mexico built entry arches and parklands at its northern border crossings. These cars are in Mexico waiting to enter the USA

Mexico's economy (Brown 1997: 118). A similar example exists on the US–Canada border where tourism has been the impetus for the opening up of several new border crossings this century. At one point on this boundary, petitions by residents on both sides of the border finally paid off when a small port of entry was changed from a 12-hour gateway to a 24-hour port. This ended years of frustration for people traveling to the neighboring country, because before they had to be home on their respective sides of the frontier before 6 p.m. most of the year in order to make it before the border gates closed. In this event, locals were required to drive a 200-kilometer detour to the next crossing point (Brooke 1996). This recent change was an important event in a region where cross-border ties are strong and where residents of small communities on both sides of the border regularly shop, dine out, and visit friends and relatives in the neighboring country.

Population changes

Although border regions have traditionally been viewed as peripheral, in some areas they are beginning to be characterized by rapid population growth and greater levels of development attention. Mexican border towns have grown more rapidly in population than their American counterparts. Between 1940 and 1980, the population of El Paso (USA) increased five times, while Juárez (Mexico), the adjacent city, grew by 25 times during the same period. From 1940 to 1980, the four US border states more than tripled in population (from 14 million to 48 million) but the six Mexican border states grew eight times, from one million to eight million (Brown 1997). By 1990 every Mexican border city was larger than its American twin, including Tijuana–San Diego, and this imbalance continues to grow (Brown 1997).

It is common for people from the Mexican interior to migrate to the border in search of work in the manufacturing plants or in tourism (Gormsen 1979). Some are successful in finding work in the *maquiladoras*, while many migrants turn to tourism as a means of support. Because it lies adjacent to the United States, the Mexican frontier zone is viewed as a land of opportunity for poorer immigrants from the Mexican interior as well as from other central American nations. This is why

> Tijuana is a magnet – for years it has been the most active illegal crossing area in all 1,936 miles of the Mexico–US border. But many of the people who come here end up staying. For them, Tijuana is a place of fresh chance, flexible class structures, hustle and hope.
>
> (Parfit 1996: 96)

Summary

As this chapter demonstrates, borders sometimes create distinct patterns in the way tourism grows and develops in border regions and within countries at large. Often, the distinctions between levels of tourism development are noticeable as soon as a frontier is crossed. Cultural and socio-economic factors, such as settlement patterns and urban structure play an important role in this. Likewise, differences can be explained by examining planning traditions, opposing ideological systems, and zoning and land use policies.

In borderlands where the divide is a major attraction, as mentioned in chapter 3, unique tourism landscapes typically come about in direct relation to the nearest border crossings. Tourist districts along the US–Mexico border are a good example. In instances where the border is merely a line of transit, a transit landscape develops. In these places the setting is characterized by a profusion of currency exchange stands, insurance dealers, restaurants, gas stations, and tourist information centers. Just as borders influence the tourism scene, tourism does in many instances contribute to the creation of distinct border environments. The demands placed by tourism on border regions commonly result in changes to local infrastructure and to a large influx of migrants who sell their possessions and move to the frontier in search of employment. This is particularly the case in the less developed countries of the world.

5 Global transformations

> In an era of international peace and cooperation, the issues of significance in a borderland are no longer drawn on national lines, but more on local and regional interests *across* the boundary
>
> (Minghi 1991: 28)

Introduction

The end of the cold war brought about a series of political events unrivaled by any in recent history. Communist rule in Eastern Europe and in countries of Africa and Latin America has fallen, giving way to the growth of democratic systems, creating in most instances conditions of friendlier relations and cooperation. Once communism had lost its centripetal power in the early 1990s, the superpower Soviet Union disintegrated relatively peacefully into 15 independent nations. Subsequent to this collapse, other eastern nations began to divide, creating new independent states on the political map of Europe. Yugoslavia crumbled, leaving five countries at war over ethnic territoriality, and in 1993 Czechoslovakia too was divided into the Czech and Slovak Republics. Such movements were not confined to Europe, for in Africa, Namibia's independence was finally achieved peacefully in 1990 after a lengthy rule by South Africa. Furthermore, South Africa's new political restructuring in 1994 finally allowed all of its citizens to participate in democratic elections and to seek public office, education, and employment regardless of race. This change also brought about the integration of the former quasi-independent black homelands into mainstream South African society.

Reunification of divided countries has also been notable during the past ten years. Britain's lease on Hong Kong expired, resulting in the reversion of the colony to Chinese administration in 1997. Likewise, Portuguese Macau was returned to China in December 1999. After decades of separation and isolation, East and West Germany were united – an act that became a symbol to much of the world that old political barriers and administrative antagonisms can be overcome. North and South Yemen were also fused into one country in 1990 despite years of opposing ideological systems (the capitalist North and the Marxist South) and strong contention on both sides.

Peace and understanding have taken over many situations where turmoil has been the norm for so long. A peaceful resolution is now being sought by some local residents and advocacy groups in the Cyprus conflict. Strife has ceased in Mozambique between government leaders and rebel guerrillas to the extent that a once terribly scarred nation is on the mend. Similar fragile peace has recently been established in Lebanon, El Salvador, and Nicaragua after years of bloody civil wars. In all of these cases, tourism is beginning to develop with a great deal of international interest. Relations between long-time rivals Israel and Jordan have improved to the extent that border crossings have become routine for tourists, Israelis, and Jordanians, and additional crossing points have been opened and are being planned. Likewise, after decades of animosity, Chinese and Vietnamese relations have so improved that cross-border exchanges between officials and tourists in both countries have recently found success.

These changes, coupled with people's improved standard of living and rapid advances in technology and transportation, have created a smaller world where few places remain unaffected by global changes in politics, economics, and tourism. Nonetheless, not all changes have been positive, nor have they all benefited the countries involved. For example, the independence of Eritrea from Ethiopia in 1993 cut off all coastal access to the rest of Ethiopia and has since resulted in territorial disputes between the two countries, which have all but eliminated any attempt by either country to develop tourism.

This chapter aims to examine some of the many geopolitical changes that have occurred in recent years, which have affected the growth and development of tourism and altered significantly the traditional roles of political boundaries.

Improved international relations

As demonstrated above, global political and economic changes abound. In many cases, antagonistic views of neighbors and strained relations have given way to friendship and cooperation (Minghi 1991). Several factors can be identified as being responsible for these changes. First, fiscal forces and economic rationale have driven the improvement of international relations and the breakdown of traditional travel restrictions between some countries. For example, some states (e.g. Indonesia prior to the 1999–2000 conflicts in East Timor) have reduced visa requirements for several nationalities, partly in an attempt to attract increased numbers of tourists and international trade. In 1997, in an attempt to attract more Chinese tourists, South Korea announced its intentions to simplify visa procedures for tourists from China. Plans include a reduction in the number of documents required for issuance of a visa and the simplification of entry and exit procedures at the border (China Radio International 1997).

Second, some nations are motivated by pressure from international organizations and advocacy groups. Several organizations have for a long time advocated the dismantling of political frontiers or at least a decrease in the barrier effects of international boundaries (World Tourism Organization 1988). In 1973, the Customs Cooperation Council (CCC) established the Kyoto Convention which contains recommendations for the implementation of several standards and practices on customs regulations applicable to travelers. For instance, the Convention recommends a dual channel voluntary system of customs declaration by travelers to facilitate more efficient and less burdensome inspections (Ascher 1984). This has been implemented in many places where it is common to see 'green' lanes (nothing to declare) and 'red' lanes (goods to declare) at international ports of entry. As a result of this and several other conventions, the WTO in 1999 established a Global Code of Ethics for Tourism wherein Article 8, Principle 1 declares the following:

> Tourists and visitors should benefit, in compliance with international law and national legislation from the liberty to move within their countries and from one State to another, in accordance with Article 13 of the Universal Declaration of Human Rights; they should have access to places of transit and stay and to tourism and cultural sites without being subject to excessive formalities or discrimination.
>
> (World Tourism Organization 2000b)

Principle 4 of the same Article recommends that

> Administrative procedures relating to border crossings whether they fall within the competence of States or result from international agreements, such as visas or health and customs formalities, should be adapted, so far as possible, so as to facilitate the maximum freedom of travel and widespread access to international tourism; agreements between groups of countries to harmonize and simplify these procedures should be encouraged; specific taxes and levies penalizing the tourism industry and undermining its competitiveness should be gradually phased out or corrected.
>
> (World Tourism Organization 2000b)

The Organization for Economic Cooperation and Development (OECD) has also fought to ease customs and immigration procedures at international crossing points. The World Tourism Organization works alongside the OECD and CCC in this endeavor. As part of its efforts, the WTO established a committee to oversee measures to simplify entry and exit formalities, to report on existing government requirements or practices that may impede the development of international travel, and to develop standards to be adopted by member states (Ascher 1984: 8). More recently, the Travel and Tourism Government Affairs Council called on governments to promote open border policies that will allow people to travel freely. In a resolution adopted by the Council, the group pleaded with all nations to ease restrictions on travel to other countries except 'under the most extraordinary circumstances' (Dorsey 1990: 6).

Liberalization of border restrictions is a slow, arduous, and uncertain process, and the large number of international organizations with jurisdiction over limited components of tourism, as well as each country's prejudices and policies, typically leads to a fragmented, uncoordinated approach to decreasing the obstacles to international travel (Ascher 1984: 9). Nevertheless, the World Travel and Tourism Council (1991) reported that several border-related barriers to travel in Southeast Asia and Europe have begun to decline, although the Council called for a more serious examination of travel barriers between nations because current efforts were insufficient.

The third factor in this pattern of global change is the fall of communism in Europe. Hall (1990a; 1990c) outlined many of the problems facing communist leadership during the early 1990s, which led to the disintegration of the Marxist system in Eastern Europe and the Soviet Union. Ethnic tension was a major divisive factor in the fall of the Soviet Union, for growth of the nation's Islamic population and rising fundamentalism fuelled by supporters across the border in Afghanistan and Iran increased intensity of internal conflict. Irredentist movements in Eastern Europe, particularly in Yugoslavia, Hungary, and Romania posed major external challenges to the communist regime. Environmental advocacy groups added to the pressure by campaigning for improved ecological conditions in mining, forestry, and heavily-industrialized regions. These, together with an increasing tendency towards western models of economic development by Moscow in the late 1980s, and the increasing laissez faire attitude towards Eastern Europe by the Soviets, all contributed to economic and social reform and eventually the fall of communism in Europe (Hall 1990c). Ripple effects from this momentous event were felt in other communist countries like Cuba and Mongolia, whose economies were formerly heavily subsidized, and whose tourist market was primarily from the Soviet Union and Eastern Europe.

Fourth is a decrease in the barrier effects of borders. Helle (1979) noted that as relations between the Soviet Union and Finland improved during the 1970s, the barrier effect of the border gradually became less pronounced. It became much easier for Finns to cross into the USSR, and cross-border tourism between the two countries increased dramatically. This became even more apparent when the border opened up to tourism as a consequence of perestroika. As Paasi (1994: 109) suggests,

> the opening of the border came at an opportune moment, for there are still about 180,000 people alive who were born there [the ceded terri-

> tories of Karelia], and an immediate boom in nostalgic journeys to Karelia ensued, with a total of 1.26 million crossings of the Finnish–Russian border in 1991–1992.

Similarly, despite the fact that China and Vietnam have long been foes, relations have improved dramatically in recent years. Their common boundary, which was sealed until 1992, is now a bustling region where commerce, trade, and even some forms of tourism (e.g. shopping and prostitution) are beginning to develop. The two governments have designated several border areas as places where people can cross for one-day trips, and talks are under way to extend the time they are allowed to stay in the other country. Leaders from both countries have been laboring to improve relationships, and they are even making progress in solving ongoing border disputes (Barr 1997).

Finally, some changes happen simply because cross-border relations have emended. For instance, South Africa made new friends by abolishing apartheid policies and initiating a democratic system. As a result, embargoes that limited international travel to and from the country for so long were lifted, and South African passports are now recognized and accepted by many governments who had earlier rejected them.

Another prominent case is the People's Republic of China and Taiwan. Years of separation and hostility between the two governments are gradually giving way to contact and cooperation. Yu (1997) describes the evolution of political relations between PR China and Taiwan and links them to changes in travel restrictions between the two entities. During the early stage of cross-Straits travel (1979–87), many people from Taiwan began visiting the mainland, even though a complete intolerance policy was still in place in Taiwan regarding travel to the mainland. To avoid the restrictions imposed by their home government, residents of Taiwan simply traveled via a third country, such as Hong Kong or Singapore, and PR Chinese immigration authorities did not stamp their passports, thus leaving no evidence of a visit. Official inter-Chinese travel began with Taiwan's decision to lift its total ban on travel to the mainland in 1987. At first, only certain segments of society were allowed to travel to PR China to visit friends and relatives whom they had not seen for 38 years. Leisure-based tourism and direct business contacts were still strictly prohibited. During the next two years, Taiwan's policies for allowing mainland Chinese to visit the island were also relaxed. The current period (1994–present) is characterized by increased travel across the Taiwan Straits in both directions, although it was temporarily restricted by the Taiwan government in 1994, when a group of 24 Taiwan tourists was ambushed and murdered in the PRC. The ban was lifted only a few months later, and since that time, tourism from Taiwan has become a major component of China's tourism industry, accounting for 19 percent of the country's total tourism receipts in 1995, and in 1997, direct sea links were opened for trade (Yu 1997). Travel is still via a third country, and now that Hong Kong belongs to China, a new situation has been created, which has yet to be resolved. Taiwan's strict 'three nos' policy – no direct postal communication, no direct air service, and no direct trade – remain the greatest barriers to greater levels of cooperation and cross-Straits tourism (Zhang 1993).

Even North and South Korea, which have been in complete isolation from each other since the 1950s, have begun to soften their defiant attitudes. Several efforts have been initiated to begin bilateral talks, but most of these have failed. In 1988, South Korea, in a surprise declaration, announced that it would permit the trade of merchandise with North Korea. The North reacted by inviting South Korean business people to Pyongyang, which resulted in limited visits by Southern business people and trade in the private sector in early 1989 (Kim and Crompton 1990).

Tourism can be credited with part of these emended relationships as it was a catalyst for change. While the basis for the opening up of Eastern Europe can be attributed to the openness created by perestroika and glastnost (Buckley and Witt 1990), tourism became one tool used by many countries to continue the wave of change. Hall (1991b: 11) outlines several ways in which tourism contributed to these changes. It was:

- a means of obtaining hard currency and improving the balance of payments by admitting more Western tourists;
- a catalyst for social change by allowing closer interaction between host communities and tourists from the outside world;
- a symbol of new freedom by allowing Eastern Europe's citizens to travel freely both within and outside their own countries, albeit constrained by financial barriers at first;
- a means of improving local infrastructure by upgrading tourist facilities with or without foreign investment;
- an integral part of economic restructuring with the elimination of centralization, subsidy, and bureaucratization.

Several unique spatial and political patterns have developed with an increase in liberalized international relations that significantly affect tourism. These include stability, visa- and passport-free excursions, extended accessibility to formerly restricted areas, unification of divided states, and the growth of international economic and trade alliances.

No-passport and no-visa excursions

Improved international relations, increased stability, and the desire for tourist dollars provided an impetus for some countries to relax their frontier formalities to the extent that residents of neighboring countries could enter without visas, and in some cases even passports. In the 1960s Anglo-French cross-channel travel grew rapidly, which caused both governments to consider the best ways to manage immigration procedures at sea ports. One of the most significant types of cross-channel traffic was excursionist travel, defined as a person who crossed the Channel for less than 60 hours, most often associated with shopping and sightseeing in port communities adjacent to the seashore. Good relations between Great Britain and France led to an arrangement that allowed British citizens to cross to France for short excursions and French citizens to Britain using only unofficial identity cards, which were issued by travel agencies and cross-channel transportation companies without any kind of background or citizenship checks (Gibb 1985). Notwithstanding their economic importance in the border region, these no-passport excursions were abandoned by the French government in 1984 owing to concerns that illegal immigration was occurring through Britain and that many British nationals were staying too long in France since the identity cards were a virtually uncontrolled means of entry. Later in 1984, after intense lobbying by public and private interest groups, the two governments reached a compromise. A new form of visitors card was issued, which enabled British residents to visit France for periods under 60 hours. The new system, however, was more bureaucratic and sophisticated than the one it replaced – evidence of British citizenship and an official application from the post office were required, it was only valid for one month, and cost £2 (Gibb 1985: 95).

In 1972 passport and foreign currency limitations between Poland and East Germany were abolished, which resulted in immediate growth in cross-border tourism. In 1971, 734,000 people crossed the border. In 1972, after the abolition, an astonishing 16.2 million people crossed the border (Warszynska and Jackowski 1979). Similarly, in the late 1980s, relations between the USSR and Finland were in such a condition that Finns were permitted to take day trips to Tallinn and Leningrad without visas, as long as they traveled in organized groups. This too created a significant demand for tourism to these 'exotic' destinations.

During the short period between the fall of the Berlin Wall and reunification of the two Germanies (approximately one year), visa restrictions between East and West were relaxed and eventually lifted altogether. Amid the jubilation on both sides of the border, Easterners were permitted to travel to the West and Westerners to the East visa-free, and thousands took advantage of this new situation (Kiefer 1989). According to Seibert (1990), with border checkpoints eliminated, traveling between East and West Berlin was made much easier, and a bicultural life was created as tourists were able to cross from side to side unhindered.

In 1995 efforts were under way in Washington, DC, to eliminate customs and immigration checks on the US–Canadian border. The proposal carried by a lobby group representing customs and immigration employees promoted the concept of an open border, which would be as easy as crossing between states or provinces in North America. According to Reza (1995: A3), this 'radical proposal by US federal agencies would let millions of Canadians enter the United States through airports and border crossings without immigration and customs inspections,' although Canadian authorities were not as receptive of the idea for protectionist reasons (*Denver Post* 1995).

Increased access to restricted areas

Owing to economic necessity and in some cases a lack of ability to keep ever more curious tourists away, governments are allowing tourism to develop in areas that have heretofore been heavily restricted from foreigners. North Korea has long been off limits to most foreign travelers except for a few official delegations, friendship groups, communist tourist groups, and overseas Koreans, primarily from Japan. However, in the mid-1980s this began to change as the Democratic People's Republic of Korea (DPRK) made the decision to change the world's perception of it by introducing an ambitious tourism development program. Plans included trips outside Pyongyang, the capital city, to regions of natural scenery and historic importance, 30,000 new beds, and improved travel linkages within the country and with its neighbors.

To pay off debts accumulated during the 1970s, North Korea invited German creditor banks to send mining technicians to help develop the nation's mineral resources. This unprecedented interest in joint ventures started a slow move toward openness to outside assistance and partnership (Foster-Carter 1981). In September 1984 a joint venture law was passed that allowed the establishment of projects on North Korean soil with foreign partners (Hall 1986a). This resulted in a joint DPRK–French project to construct a 40-story hotel in Pyongyang, although the project eventually failed owing to bad planning and surveying on the part of the Koreans. The second venture involved Japanese partners in establishing a chain of hard currency shops, Ragwon, where goods not normally found in North Korean shops could be purchased with foreign currencies (Hall 1986a).

More recently North Korea hosted a conference aimed at luring foreign companies to invest in a new free trade and economic zone along the border with China and Russia in 1996. This is a significant move toward capitalism, which is expected to be limited to the Rajin-Sonbong Free Economic and Trade Zone. The area has a natural harbor that could serve as a gateway to other parts of Asia. This move falls in line with the 1984 joint venture law described earlier. Likewise, on June 9, 1997, North Korea decided to designate Pidan Island, in the Yalu River along the border with China, as a free economic trade zone that ensures investments and free visits. North Korea plans to develop the 70-square kilometer island into a city of manufacturing, high-tech industries, international trade, entertainment, and tourism (Ul-Chul 1997).

Tibet is another area that has long been considered off limits to foreigners. Since the beginning of 1994 the Tibet Autonomous Region has been opening up and establishing links with foreign countries and developing border trade and tourism. Several border crossings were opened in 1994 with India and Nepal, which marks the end of a 30-year trade suspension with India (*Beijing Review* 1994). During the early 1990s Tibet opened up rapidly to tourism, and now over 23 new international tourist hotels have been built and more than 20 travel agencies in the area have been created.

Many areas of the Soviet Union were off limits to tourists for many years, including the industrial communities near the Finnish border. Since the collapse of the USSR, however, it has become easier for Finnish nationals to travel around in the ceded areas, and thousands have participated in nostalgic tours to the region (Paasi 1996; Paasi and Raivo 1998).

As discussed in chapter 2, Albania, which was until fairly recently one of the most isolated countries in Europe and the world, used tight controls to restrict access to certain parts of the country, and the country as a whole, to most foreigners. Individual tourist visas were nonexistent, instead group visas were required. These were time consuming and required a great deal of forethought and planning. Tourists were required to enter the country as a group and travel together as a coherent group at all times. Contact with local residents was kept to a minimum, and Albanians were required to report any personal contact with tourists. Itineraries, accommodations, guides, food, and transportation were all strictly controlled by Albturist. Foreign guidebooks about Albania were banned, and a person's physical appearance (e.g. long hair and unshaved face) was grounds enough to be refused entry into the country (Hall 1984; 1992). This near-paranoid approach to tourism has since given way to open arms policies and international promotion. A Ministry of Tourism has been installed to encourage inbound tourism, focusing on the country's coastline, historical sites, and mountain scenery (Witt 1998: 386).

A different perspective was that of the West by residents of the East. From the point of view of Eastern Europeans, access to the West was, until the 1990s, heavily restricted, with a few notable exceptions like Yugoslavia. However, beginning in 1990, the forbidden West became the destination of choice for Easterners. This created a unique phenomenon in Europe where tourism from the East lagged significantly behind that from the West. As Hall (1995) suggests, Western Europeans are now interested in alternative forms of tourism, while residents of Eastern Europe are only now beginning to travel as mass tourists.

> Just when many analysts have been predicting the death of Western mass tourism with the development of niche, special-interest and 'quality' products, there is now a growing CEE [Central and Eastern Europe] market for just the types of package holidays which were popular with Western Europeans in the late 1960s–1980s.
>
> (Hall 1995: 235)

Unification of separated countries

As mentioned previously, reunification of separated states has occurred recently. Reunification is a difficult process that involves a complete re-evaluation of spatial policies (F. Smith 1994). It brings about many opportunities such as more equitable development in larger territories, but it also creates major challenges. These will be discussed in greater detail below.

Most of the consequences of the 1990 German reunification are seen in economic terms, although political and environmental problems also ensued. First, thousands of state-run companies in East Germany, including hotels, restaurants, tour companies, and shopping facilities, had to be privatized. This resulted in severe dives in employment rates to the extent that over 1.5 million jobs that were previously guaranteed by the communist government disappeared (Mo 1994: 54). Finding and creating jobs for the unemployed, particularly those who were formerly employed by the state (e.g. as East–West border guards) created a terrible strain on the country.

Second, the integration of prices, social security benefits, and wage levels proved to be a difficult task. The economic systems of the two Germanies had developed in complete isolation from each other, with the result that necessities such as food, housing, public transportation, newspapers, and education were highly subsidized in the East, and luxuries like televisions, washing machines, and cars, were expensive and difficult to acquire. West Germany, however, had developed into one of the world's most prosperous countries where private enterprise was encouraged, and the role of government was much smaller in economic affairs (Roberts 1991: 380).

Third, as privatization began people started to claim family land and property that had been seized from them after the Second World War. Property rights questions had to be settled by the courts, but the judges who previously worked for the East German state were viewed as being incapable of making legitimate decisions in a new democratic society. Thus, many were released from duties and reassigned to other positions, while other judges, like many Eastern civil servants underwent extensive retraining (Roberts 1991).

Fourth, the Democratic Republic's infrastructure was far from the level of development of the West. Billions of Deutschmarks were spent in modernization efforts (Roberts 1991). Eastern roads were too small for the demands of modern traffic, and sewers, railway networks, telephone systems, and public buildings were in primitive conditions compared to those in the West (Bucholz 1994; Harris 1991). Similarly, with privatization, heavy industries in the East required huge outside investments to modernize factories, telecommunications services, and computers (Harris 1991).

The fifth primary concern in reunification efforts was environmental issues. Safety and environmental standards in the East were much more relaxed than the rigorous regulations in West Germany (Harris 1991). The range of ecological problems included high levels of soil, water, and air pollution, from facilities as large as nuclear power plants to individual service stations, exacerbated by the use of rivers as open sewers and the uncontrolled dumping of dangerous chemicals (Roberts 1991: 381).

Sixth was the military alliances. For several months the Soviet Union argued that is was not possible for the new Germany to remain in NATO because of the East's membership in the Warsaw Pact. Eventually this problem was solved by treaty between Germany and the USSR, and eventually the whole of Germany became a member of NATO (Harris 1991; Mo 1994; Roberts 1991).

Finally, rapid immigration of easterners to the West created additional problems. Border towns of the Federal Republic became areas of immigration overload. In 1989 alone, 344,000 East Germans moved west, following which trainloads and carloads of people began arriving at border stations in the amount of 1,500 to 3,000 a day (Mo 1994). This resulted in a 30–50 percent increase in new registrations for unemployment benefits in October and November 1989 (F. Smith 1994), putting a strain on the West German states nearest the border.

Some observers have suggested that the costs of reunification were amplified because the event happened too quickly. Unlike other East European countries, the economy of East Germany was integrated into the world economy too fast without an adjustment period in which the two sides could negotiate economic and political conditions (Mo 1994).

> Reunification, like division, alters the geographical incidence of peripherality. Economic readjustment is just as problematic as at the time of division and currently involves more change, particularly on the eastern side, because of the gap between East and West in economic terms as a result of the past 40 years.
>
> (F. Smith 1994: 238)

Efforts are well underway to bring to fruition the reunification of North and South Korea. Observers believe that Korean unification is a realistic possibility for the near future (Eberstadt 1995) and that the new Korea will face similar problems. Leaders are looking at the German model to prepare for the momentous occasion (Henriksen and Lho 1994; Mo 1994). Kim and Crompton (1990: 156) described four interrelated forces that provide pressure for the unification of the Koreas. First, the level of Korean ethnic homogeneity is matched by very few nations elsewhere in the world, and it is suggested that nearly ten million families have been separated between the two sides on the Korean Peninsula. These divided families exert a great deal of pressure for reunification. Second, it makes economic sense. Billions of dollars are spent by both Koreas every year to maintain military power along the demilitarized zone. Most of these financial resources could be used for more productive and peaceful ends if both sides were united.

Third, leaders of both Koreas have emphasized their goal of a united nation for decades, even if their approach to unification is drastically different. Fourth, reunification would contribute to the maintenance of global peace.

The reunification of the Koreas could potentially be more problematic than the German experience (Foster-Carter 1994). According to Kim Kyung Won, South Korea's former ambassador to the UN, the two Koreas have radically different perceptions of the process of reunification. The South, in his opinion, is preoccupied with a peaceful process such as Germany's, while the North wants a unification achieved through conquest, such as in the case of Vietnam (cited in Hoon 1990: 11). Kim and Crompton (1990) argue that tourism has the potential to assist in the reunification process of the two Koreas because it is a means of reconciliation and peace. In the Koreas, tourism-related projects have been proposed during the past 15 years to increase levels of benevolence between the two sides. For example, the South suggested opening the Seoul–Pyongyang Highway; postal exchanges; reunions of separated families; exchanges of goodwill visits by people with similar interests; exchanges of sports teams and participation in international games with a single, unified delegation; and creation of sports facilities and parklands within the Demilitarized Zone (DMZ) (Kim and Crompton 1990: 361). In 1985 limited groups of ordinary citizens and artists were allowed to travel to the other side. For the first time in years people were permitted to cross the border, although this experiment was short-lived. Tourism between South and North Korea is beginning to develop, however, comprising of South Koreans and overseas visitors. Tours from Seoul take groups to Panmunjom, the armistice village on the border, and boats are now shuttling foreign visitors around the DMZ to areas like Mount Kumkang (*Chosunilbo* 1999).

Another form of unification is when a colonial territory is absorbed by the neighboring country. These events differ somewhat from two countries becoming one entity because territorial sovereignty is transferred from one state to another, not two sovereign states becoming one.

The British return of Hong Kong to China is July 1997 is a good example of this. Prior to this several observers speculated on the types of problems that might occur when sovereignty changed hands from a capitalist European power to the hands of communist rulers. Particular attention has been paid to how the lives of everyday Hong Kong residents would be changed, how business and international trade would continue, and how tourism would proceed to function as it had until then. Chinese authorities have long insisted that they would employ a 'one country, two systems' approach, meaning that life in capitalist Hong Kong would continue as before, separate from the rest of China as a Special Administrative Region, but functioning within a Beijing-based regime.

In the context of tourism, a number of concerns were alluded to by observers. The new airport, completed in 1998, was the center of much speculation. Arguments over the funding of the airport using British and Chinese resources created serious contention between the two countries (Hobson 1995). Similarly, it was suggested that difficulties might arise concerning the airline industry. Since China did not allow airlines that served Taiwan also to serve the mainland, Dragonair has traditionally held rights to fly to China, while Cathay Pacific has served Taiwan. This, according to Hobson (1995: 16), would affect the future of both airlines, for all traffic rights and bilateral air treaties 'will have to be agreed by China. At the stroke of a pen, these rights and treaties could be rewritten.'

Additionally, as property investors realized that monetary return on commercial properties would be higher than hotel ventures, some hotels were razed and new office buildings erected in anticipation that tourism would decline in the territory as it had in mainland China immediately following the Tiananmen Square event (Heung 1997; Hobson 1995). Concerns were also expressed about inbound and outbound travel. Prior to 1997, Taiwan was Hong Kong's most important market origin in terms of both arrivals and expenditures, and observers worried that with Hong Kong becoming part of China, some kind of change would occur, particularly the likelihood of formal or informal direct air links between China and Taiwan. However, if political tensions increased then Taiwan could prevent its citizens from traveling to China through Hong Kong since the latter will have become part of China (Hobson 1995: 17). In terms of outbound travel, questions of passport possession were at the forefront together with the rights of Hong Kong's residents to travel abroad. British National Overseas (BNO) passports, which were to supercede the British Dependent Territory Citizen passport, or the Special Administrative Region (SAR) passport issued by China were sure to cause problems when traveling abroad, particularly since visa-free travel on BNO and SAR passports had been arranged with only a few countries (Hobson 1995).

Mok and Dewald (1999) provide a useful overview of tourism in Hong Kong after the handover to China. In July 1997, immediately following the transfer of sovereignty, tourist arrivals in Hong Kong experienced a drastic decline, amounting to a 35 percent drop from July the previous year. For all of 1997, visitor arrivals were down more than ten percent, which resulted in a loss of millions of dollars in revenue, and thousands of lost jobs in the tourism industry. Several reasons for this rapid decline have been identified (Mok and Dewald 1999):

- Travelers' fears and lack of knowledge about what would transpire after the handover was a factor in this decline, even though China's government had promised the world that life would carry on just as before.
- Negative reports in foreign media created sensational images of what could go wrong, often citing the 1989 Tiananmen Square incident. Several American newspapers ran headlines that struck fear and uncertainty in the minds of potential foreign travelers.

- The Chinese government introduced tighter restrictions on Chinese, traditionally one of Hong Kong's largest market segments, entering the new SAR. This move on the part of the Beijing government was a demonstration that it intended to alleviate local fears that masses of people from China would enter Hong Kong after July 1 and that it intended to protect the ex-colony's status as a SAR.
- When Hong Kong became part of China, Taiwan's restrictive laws were still in place, which required Taiwanese tourists (another of Hong Kong's largest market segments) to undergo complicated immigration procedures and travel to a third party country. This clearly contributed to an immediate decline in arrivals because many Taiwan residents preferred to enter China via Macau.

Despite these initial difficulties, tourism in Hong Kong has started to mend, and as promised before the changeover, China has maintained minimal interference in the SAR's politics, and in terms of tourism administration, everything has been business and life as usual. According to one local corporate executive, 'there is nothing changed in Hong Kong after the handover except the flag at the top of buildings' (quoted in Mok and Dewald 1999: 36). Beijing has promised similar treatment of the Portuguese colony, Macau, following its return to Chinese rule in December 1999, and owing to the importance of tourism in the colony, it will likely be enhanced and coddled by China. In fact, some argue that this transition will be even easier because Hong Kong's transition period was rocked with poor Sino–British relations, while nothing but goodwill between China and Portugal has prevailed over Macau's transfer of sovereignty (Porter 1997).

International economic alliances

Nations today realize that the isolationist view of sovereignty and independence no longer works the way it has for centuries. Nearly every aspect of domestic policy has some kind of connection to the outside world, which is creating the need for states to learn to deal with not only the international effects of globalization, but the domestic effects as well. This results in the need for strategic alliances among states on various regional scales 'to secure the basis for economic and political survival as the imperatives of structural competitiveness make themselves felt' (Jessop 1995: 682).

Dozens of international regional alliances have been formed during the twentieth century, most focusing on economic development or military partnerships (see Table 5.1). While military pacts are important in a regional context, this section focuses primarily on trade blocs and economic alliances that have significant implications for tourism. The primary theory behind economic integration is that by combining their economies two or more countries can reduce production costs by producing for a larger mar-

Table 5.1 A selection of regional political and economic alliances (past, existing, and proposed)

Arab Common Market
Arab Maghreb Union
Association of Southeast Asian Nations (ASEAN)
Australia–New Zealand Closer Economic Relations Trade Agreement (CER)
Asia-Pacific Economic Cooperation (APEC)
Black Sea Economic Cooperation
Caribbean Community (CARICOM)
Central African Customs and Economic Union (UDEAC)
Central American Common Market (CACM)
Common Market for Eastern and Southern Africa (COMESA)
Commonwealth of Independent States (CIS)
Council of the Baltic Sea States
Economic Community of Central African States (CEEAC)
Economic Community of West African States (ECOWAS)
European Free Trade Area (EFTA)
European Union (EU)
Latin American Association for Development and Integration (ALADI)
Latin American Economic System (SELA)
Nordic Council
North American Free Trade Agreement (NAFTA)
Organization of African Unity (OAU)
Organization of American States (OAS)
South Asian Association for Regional Cooperation (SAARC)
South Pacific Commission (SPC)
South Pacific Regional Trade and Economic Cooperation Agreement (SPARTECA)
Southern African Customs Union (SACU)
Southern African Development Community (SADC)
Southern Common Market (MERCOSUR)

ket base; producing a wider variety of goods; and having a louder voice in economic and political affairs (Glassner 1996: 411). According to Glassner (1996: 411–12), there are four levels, or degrees, of economic integration. First, free trade areas exist where two or more countries agree to eliminate tariffs and trade barriers so that goods flow freely between them. Each member, though, retains its own trade restrictions with other non-member parties. Second, a customs union comprises a free trade area plus a common external tariff, so that one unit is formed for trading with non-members. Third, common markets are essentially customs unions that allow the free movement of capital, labor, and goods. Any citizen of any member country can be employed in any of the member countries, and all boundaries are eliminated in terms of investments and banking. Finally, economic unions exist when complete economic integration is achieved. Members have a common market plus a common currency, common monetary policy, uniform banking and insurance systems, and common taxes and corporation laws.

The North American Free Trade Agreement and the Canada–US Free Trade Agreement

There is a growing literature on the growth, development, and economic impacts of bi- and tri-lateral economic associations in North America (e.g. Grinspun 1993; Hocking and McGuire 1999; McMillan 1993; Randall and Conrad 1995; Richardson 1993). In January 1989, the Canada–US Free Trade Agreement (FTA) was established to begin eliminating trade barriers between the two countries in the area of goods and services. This agreement was a milestone for tourism in North America because it addressed the industry directly and in detail. The document's tourism annex included the following allowances:

- the adoption and fees on the departure or arrival of tourists would be determined by each nation and would be limited to the approximate cost of the service provided;
- the promotion of tourism in the territory of the other country including joint efforts by national, provincial, state, or local governments;
- yearly consultation to identify and strive to eliminate barriers to trade in tourism services;
- identification of ways to increase tourism between the two parties;
- no restrictions on the value of travel services that may be purchased in the other country (Taylor 1994: 316).

In 1989 and 1990 other bilateral agreements were signed between Canada and Mexico and the United States and Mexico that dealt directly with tourism-related issues. These included the facilitation of travel between countries, promotion of cultural exchange, cooperation in human resource training, joint marketing activities, and assisting each other in achieving WTO uniform standards and practices (Taylor 1994: 316).

In 1994 the FTA was superceded by the North American Free Trade Agreement (NAFTA), whose primary goal was extended to eliminate tariffs and other trade barriers between Canada, the United States, and Mexico gradually over the next 15 years. Tourism was not addressed explicitly in this treaty as it was in previous agreements, but it is covered indirectly under the concepts of temporary entry, trade in services, financial investments, and telecommunications development. According to G. Smith (1994), this is the strength of NAFTA, that unlike the earlier FTA, all services are covered by the agreement unless specifically excluded.

The European Union

A great deal of attention has also been directed at the European Union in recent academic literature (Cole and Cole 1993; El-Agraa 1990; Sweet and Sandholtz 1997), particularly related to economic policies, peripheral development, trade, and political boundaries. From its foundation in the 1950s as the European Economic Community, the Union has gone through several changes, which have been effective in eliminating many trade and human barriers in Western Europe. The Union's original goals were wholly economic in nature, but have since expanded to include foreign policy, citizenship, and security arrangements. Since its foundation, the Union's goals have leaned towards supranational arrangements in economic and political terms. To create a frontier-free region in Western Europe, the Schengen Agreement was signed in 1990 by France, Germany, Luxembourg, Belgium, and the Netherlands, followed later by Portugal, Spain, and Italy. The goals of the 1992 Maastricht Treaty were manifold (*Economist* 1991). However, perhaps the most important for the present discussion of tourism was the complete abolition of immigration and customs controls at borders inside 'Schengenland' and initial talks on common visa policies. Although the Schengen Agreement aimed to create a completely frontier-less region, it took several years for the total removal of border controls between signatories (Cole and Cole 1993; Cowell 1997; Huntoon 1998). The Maastricht Treaty reinforced the Schengen goal when the EC was officially transformed into the European Union (1 January 1993). This designation officially eliminated customs, immigration, and trade barriers between member states and created the single market, which guaranteed free movement of people, goods, capital, and services within the EU.

As of 1999 EU membership included Austria, Belgium, Denmark, Finland, France, Germany, Greece, Ireland, Italy, Luxembourg, Netherlands, Portugal, Spain, Sweden, and the United Kingdom. Other European nations, primarily from the former Warsaw Pact, are vying for membership in the EU. Talks are currently under way to begin entry negotiations for Poland, Hungary, Slovenia, Cyprus, Estonia, and the Czech Republic. Plans are also being made for discussing membership for Latvia, Lithuania, Romania, Bulgaria, and Slovakia (Echikson 1997).

For over a decade the European Commission – the governing body of the EU – has worked to promote social and economic cohesion and reduce regional disparities. This has often included tourism. However, for the first time, the industry was repositioned in the EU policy framework with its recognition as a separate entity in the 1992 Maastricht Treaty on European Union (Williams and Shaw 1998). Because the European Union is perhaps the most truly integrated of all international regional alliances, it has demonstrated the greatest effort to develop tourism policies. So far many studies have been conducted, policies formulated, and development projects undertaken specifically with tourism as the object of attention. Structural, administrative, and financial support from the Commission has been strong in these areas and continues to be a integral part of the Commission's efforts.

The Association of Southeast Asian Nations

The Association of Southeast Asian Nations (ASEAN) began in 1967 in response to the threat of communism in the region. The member nations at that time feared that they might be overrun by communist forces, so they united in an effort to avoid foreign occupation (Hagiwara 1973; Rimmer 1994; Tasker *et al.* 1994). After the withdrawal of US forces in the mid-1970s and the establishment of communist governments in the region, ASEAN member nations recommitted themselves to work together and protect each other. In 1976 the first ASEAN summit was held in Bali, Indonesia, where economics dominated the discussion (Broinowski 1982; Hussey 1991), and a treaty was signed to strengthen economic cooperation and mutual assistance (Castro 1982). This time was critical in ASEAN history, for this is when political concerns were superceded by economic terms. Membership has grown to include Brunei, Cambodia, Indonesia, Laos, Malaysia, Myanmar, Philippines, Singapore, Thailand, and Vietnam, and monumental efforts are under way to develop cooperative relationships in terms of production and trade in goods and services.

While the original treaty completely ignored tourism, the 1976 agreement stated that member states shall take cooperative action in their national and regional development programs to broaden the complementarity of their respective economies. This, in common with so many of the world's economic alliances, laid the groundwork for additional talks of intraregional cooperation for tourism. In 1987 the Manila Declaration declared that intra-ASEAN travel would be encouraged among member nations and that a competitive and viable tourism industry should be developed. In 1992 member states agreed to increase cooperation in tourism promotion, and in 1995 the Bangkok Summit Declaration reiterated the need for greater networking in all service areas, including tourism, and set forth the notion of promoting sustainable tourism, preservation of cultural and environmental resources, the provision of transportation and other infrastructure, simplification of immigration procedures, and human resource development.

Only recently, however, has ASEAN begun to consider the effects of tourism in earnest. Member states recommitted themselves to work together in the tourism arena in 1998 by formulating the Ministerial Understanding on ASEAN Cooperation in Tourism and bypassing the Plan of Action on ASEAN Cooperation in Tourism. While most areas of planned collaboration thus far have dealt primarily with marketing and promotion (Timothy 2000c), efforts are currently under way to examine broader environmental and border-related tourism issues as well. Meetings in 1999 between tourism ministers of all ten members reaffirmed their commitment to work together on a regional basis for tourism.

Few ASEAN-wide efforts have so far been undertaken or policies developed in relation to tourism. The recency of the industry as a consideration in ASEAN negotiations has resulted in policy issues and implication of tourism having been put in the form of goals and objectives for the future. Goals and strategies established in the two 1998 documents include: (1) marketing ASEAN as a single destination; (2) encouraging tourism investments; (3) developing a critical pool of tourism human resources; (4) promoting environmentally sustainable tourism; (5) facilitating seamless intra-ASEAN travel; and (6) facilitating the exchange of information and experiences (ASEAN Secretariat 1999).

Australia–New Zealand closer economic relations

In common with most other economic alliances, the Closer Economic Relations (CER) agreement between Australia and New Zealand aims to bring down trade barriers between the two countries. The focus of CER is trade in goods, not services, which tends to exclude tourism from the equation, although air transportation was an integral part of the agreement and has major implications for tourism (Hall 1994a; Pearce 1995).

Other economic alliances

The trading blocs described above have clear tourism-related implications and are located in the fastest growing tourism regions in the world. Many other regional alliances exist throughout the world that may have significant bearings on tourism. The Economic Community of West African States (ECOWAS), the Southern African Development Coordination Conference (SADCC), the Preferential Trade Area (PTA) for Central, Southern, and Eastern Africa (Dieke 1998), the Southern Common Market (MERCOSUR), the Latin American Free Trade Association (LAFTA) (Cárdenas 1992), the South Asian Association for Regional Co-operation (SAARC), and the Common Market for Eastern and Southern Africa (COMESA) are only a few examples. Other alliances are currently being designed and implemented as the 'new countries' of the world are beginning to realize the potential value of establishing free trade areas. Estonia, Latvia, and Lithuania, for example, recently signed a tri-lateral free trade agreement that will ease economic relations between them, and which may have potential impacts on tourism (Laar *et al.* 1996).

Challenges of international alliances

As regards new alliances, Jessop (1995) outlines seven challenges facing their development, particularly in the context of Eastern and Central Europe. First, the inherited administrative frameworks were poorly developed, and local authorities still have few formal competencies to pursue regional economic strategies. Second, administrative areas typically do not coincide with historical regions. Third, few intersectoral linkages were formed and maintained during the socialist period, which resulted in the emergence of a distinct sectoral or branch economy. Fourth, large enterprises had a key role in social life, which has invoked severe deficiencies in social policy. Fifth, regions are much less diversified economically compared to those in capitalist societies. Sixth, under the communist state, there was a great degree of uneven development between cities and rural areas, and border and remote regions were neglected, depopulated, and underdeveloped. Finally, the western portions of socialist economies were better developed than those in the East (Jessop 1995: 984–5).

The success of international alliances is largely dependent on the level of trust and friendship between countries. Achieving economic integration is a difficult and time-laden task, particularly when countries are reluctant to give up any degree of sovereignty and control over decision making (Glassner 1996; Timothy 2000c).

According to Kolossov and O'Loughlin (1998: 264), the future global economy will not be constrained by state boundaries because globalization processes are creating new identities. The European Union is an excellent example of this where a macro-regional identity is beginning to develop. Some people go so far as to suggest that even a macro-European heritage is beginning to develop (Ashworth 1995).

Despite the wide range of attention given to international trading blocs, relatively few researchers have examined tourism and international alliances, even though supranational economic, social, and environmental policies clearly have major implications for the development of the industry. Some alliance treaties have considered tourism within the broader economic issues, but most have not taken it directly into account. Rather, much of what is known about trading blocs and tourism is derived from treaty statements dealing with areas like transportation, environment, and the cross-border flows of people, goods, and services.

Tourism development

In nearly all cases observers believe that the growth and development of international alliances will necessarily result in the growth and development of tourism. For example, Taylor (1994) argues that the NAFTA will result in better economies in North America, which should in turn result in more pleasure and business travel between member states. In G. Smith's (1994: 324) words, 'as travel and investment increase, so too will business travel, and where business travel lead, pleasure travel will follow,' thereby resulting in an increased use of credit cards, airlines, hotels, restaurants, rental cars, tour buses, cruise ships and other tourism services. In Canada, it is believed that prices will decrease owing to the elimination of tariffs on tourism-related goods like furniture, building equipment, food, and drinks, providing yet another stimulus for tourism growth by improving the country's competitive advantage.

One of the primary goals of the European Union is to reduce regional disparities among its member nations. Aid is provided through four structural funds: the European Regional Development Fund (ERDF), the European Social Fund (ESF), the European Agricultural Guidance and Guarantee Fund Section (EAGGF), and the Financial Instrument for Fisheries Guidance (FIFG). In 1975 the ERDF was created to help correct the regional imbalances within the European Community by taking part in the development and structural adjustment of areas where development was lagging behind, particularly in declining industrial regions (Wanhill 1997). Although tourism has played a relatively minor role in the Union's overall economic development goals, some efforts have been made that have included tourism, most of which have been funded through the ERDF (Pearce 1992). The Commission's ERDF-related policy for the tourism sector has two aims: to use tourism as a means for economic development and to assist in diversification efforts in regions that are too dependent on tourism or those that are most affected the seasonal nature of tourism (Williams and Shaw 1998).

Through the ERDF and its structural funds, the EU has also initiated a series of programs that were negotiated between member states and the Commission on the basis of regional or national development plans to encourage more balanced economic and social development, thereby strengthening cohesion in the Union (Wanhill 1997: 56). As part of this broader move towards equalizing development throughout the Union, several European Community initiatives were established during the late 1980s and early 1990s to bring greater ecological, social, and economic balance to some of the more neglected parts of the region (see Table 5.2). Although most of these had some implications for tourism, as can be viewed from Table 5.2, Envireg, Interreg, and Leader are perhaps the most influential for the industry. The aim of Envireg was to assist peripheral areas of the Community in dealing with their environmental problems, particularly in coastal areas. Mitigating the environmental impacts of tourism in sensitive areas was a major goal of this program. Interreg was designed to promote cooperation between adjacent border areas within the EU, particularly in areas of employment where job losses were expected to occur as changes in customs and other border-related activities disappeared upon completion of the single market (Gonin 1994; O'Dowd *et al.* 1995). Tourism became an integral part of these initiatives as a form of alternative employment for border residents. Interreg also focused heavily on cooperation in tourism development between EU and non-EU members, such as Italy–Switzerland and France–Switzerland (Leimgruber 1998) and Germany–Poland (Scott 1998). Leader was established in 1991 to fund resourceful and integrated rural development initiatives led by community groups. Tourism too played an important role in this program and benefited from EU involvement by

Table 5.2 EU initiatives for socio-economic development

Initiative	*Year(s)*	*Goals/purpose*
Envireg	1990–93	To assist the EU's least favored regions in overcoming environmental problems.
Interreg (I and II)	1990–99	To promote cooperation between border areas and assist in overcoming special development problems related to peripheral locations.
Leader (I and II)	1991–99	To promote innovative rural development programs led by community groups.
Resider (I and II)	1988–99	To support the economic and social conversion of declining industrial areas.
Rechar (I and II)	1989–99	To support the social and economic rejuvenation of declining industrial regions.
Regis (I and II)	1992–99	To strengthen ties between the most remote parts of the EU (i.e. members' overseas territories) and the rest of the Community and to improve cooperation with neighboring non-member nations.
Prisma, Stride & Telematique	1990–99	To assist in the modernization of small to medium sized businesses in terms of business services, technology, and telecommunications services.

Source: Leimgruber 1998; Scott 1998; Wanhill 1997

being targeted for international promotional efforts, technological developments (e.g. organized reservation systems for accommodations like bed and breakfasts), production of information and literature, improved quality of services, and provision of equipment (e.g. swimming pools and tennis courts) (Wanhill, 1997: 57).

As mentioned previously, in 1992 in Maastricht tourism was officially acknowledged by the European Commission as a distinct entity within the EU's administrative framework. This recognition laid the foundation for continued activity in environmental protection and new activity in the areas of education, training, culture, and transportation (Williams and Shaw 1998).

In addition to current and former efforts, the European Commission has identified three primary areas for future EU involvement in tourism:

- supporting improvements in tourism quality by examining more closely the trends in tourism demand;
- encouraging the diversification of tourism products by supporting the competitiveness and profitability of tourism;
- supporting the principles of sustainable tourism by balancing growth and conservation (European Commission 1995; Wanhill 1997: 65).

Although the priority given to tourism by the EU is still rather low, the establishment of a Tourism Division within the Commission and the use of tourism as a development tool in peripheral regions demonstrate that more consideration is being given to tourism in Europe than in other economic alliances.

While still holding a rather peripheral role in regional integration, tourism is also being used in ASEAN as a regional development tool. Future development goals include a harmonization of tourism investment policies between members nations, developing an ASEAN tourism investment guide to help standardize the procedures and regulations in member countries, to encourage joint investment missions to third countries, to remove barriers and lengthy procedures that constrain the flow of foreign investments, facilitate the free flow of labor in tourism (ASEAN Secretariat 1999).

Investments by firms in one state will likely be assisted by global alliances as barriers are reduced for banking, insurance, and real estate investments. These feature prominently in the NAFTA, ASEAN, and EU documents. According to Article XI (Investment, Services and Related Themes) of NAFTA, investors from member nations will receive national treatment, which means that any investor will receive the same treatment within a member state as investors would from within the state. Likewise, Article XII holds that trade and the supply of services can be provided by any Mexican, Canadian, or American person or firm. From the perspective of Mexico, these points will stimulate investment in Mexico by US and Canadian corporations, which according to Rodríguez and Portales (1994: 320), may limit the growth of Mexico's diversification into other tourism markets like Europe and Asia and limit the level of control of Mexico-based companies in that country's industry.

Flow of people

One of the most common goals among international trading blocs is to decrease the barrier effects of borders for goods and people. Three groups of people in tourism tend to be most affected by this: (1) tourists from within the alliance; (2) tourists from outside the alliance; and (3) industry workers from within the alliance.

The elimination of border controls created a free movement of European and non-European citizens within the Union, which according to Tzoanos (1993), will bring about a notable growth in tourism in the future. Kearney (1992: 36) also argues that 'the phasing out of customs controls, the abolition of checks on vehicles at frontiers and the introduction of the European passport constitute major boosts to tourism not only for tourists from the member states but also for those from elsewhere.' Likewise, as mentioned earlier, one of the goals of the Maastricht Treaty was to institute a common EU visa (Huntoon 1998). This, it is believed, will remove a great deal of aggravation on the part of foreign travelers to and within the region, which will probably increase tourism to Western Europe (Robinson and Mogendorff 1993).

Under the CER agreement, discussions ensued to reduce the time and inconvenience of having to clear customs and immigration when traveling between Australia and New Zealand by the creation of a 'common border.' Recommendations were made to allow passengers from other countries to complete entry procedures at the first port of entry and departure procedures at the last call. According to observers, easing entry and exit formalities between Australia and New Zealand will increase the number of visitors into the region (Pearce 1995). Progress towards the common border idea has been stalled as a result of differences in immigration and visa policies between the two countries, issues concerning the flow of criminals and welfare recipients, the requirement for cash transaction reporting held by Australia but not New Zealand, and the question of being able to intercept prohibited goods and apply export controls (Pearce 1995: 117).

Likewise, plans in Southeast Asia include streamlining inspection processes at national borders and airports by standardizing customs, immigration, and quarantine systems and procedures between member states (ASEAN Secretariat 1999). This will likely prove to be very difficult, because of the fact that the alliance is comprised of so many different government types ranging from capitalist democracies to communist regimes to military dictatorships. G. Smith (1994) also believes that tourism in North America will thrive as cross-border travel becomes easier for residents of NAFTA member nations. (This chapter later discusses some of the difficulties being experienced on the US–Canada border.)

Not all observers agree with this simple view. Leimgruber (1998), for example, argues that the demise of immigration and customs procedures at borders is probably not as influential in the decision to travel abroad as are increased affluence and mobility, although he does acknowledge that short-term international travel (e.g. day trips) will probably increase as a result of freer transboundary mobility. In some cases the abolition of borders will create a marked decline in certain types of tourism (e.g. cross-border shopping and gambling) as the border advantage disappears (Maillat 1990).

Human resources in tourism are also affected by the implementation of the single market, being made even easier with a single currency unit. Workers from any country within the European Union are permitted to work in any other member country, although language abilities, experience, and training determine the types and locations of jobs they can acquire. Article XVI of NAFTA permits the temporary entry of people involved in trade and services into any of the three countries. Not only will this affect the level of business travel, it will probably also increase the number of tourist industry workers, such as drivers, interpreters, and managers (Rodríguez and Portales 1994). In human resource terms, the US–Canada FTA was instrumental in allowing rights and obligations related to trade in services. Chapters 14 and 15 of the treaty contain provisions regarding right of establishment of commercial presence, licensing and certification procedures for service providers, and the facilitation of cross-border travel by business people (Hufbauer and Schott 1992).

Transportation and tourism infrastructure

Transportation and tourism infrastructure is another area that will likely be affected by the development of regional trade alliances. This is currently a hot topic in Europe where the Commission is attempting to analyze and make decisions regarding the details of air alliances between European carriers and their effects on fares and routes in Europe (Belden 1997). Trans-European transportation systems are gradually being introduced, encouraged by Commission actions in regulated sectors, such as the creation of a European air space and a recent review of regulations pertaining to passenger road travel with the goal of improving international coach transport (Kearney 1992: 36). Presently the European countries operate several state-supported airlines, which have attempted to avoid the free market requirements of the single market. However, in 1986 the EC Court of Justice judged that government-run airlines were not free from integrated economic policies and must begin the deregulation process. This is expected to result in a 15 to 20 percent decrease in the cost of air travel within Europe, particularly within the early stages (Mihalik 1992: 28).

Multinational air agreements were not addressed in either the FTA or NAFTA, and it appears that they will continue to be negotiated bilaterally and separately (Taylor 1994). For example, a new air transport agreement has increased the ease with which US and Canadian carriers can enter cross-frontier markets. Canadian carriers may now conduct scheduled passenger services from any point in Canada to any point in the United States without restrictions on frequency, capacity, and aircraft size. However, restrictions on the transport of passengers to and from third countries still exist (Bailey 1995). A similar situation exists in the realm of CER. One of the most prominent themes in that agreement was aviation. According to the agreement, the two countries would be joined into one large aviation market. This meant that trans-Tasman aviation would be opened up to airlines other than Qantas and Air New Zealand. It also meant that an all-points exchange would be phased in by the end of 1994, which would allow all Australasian airlines to operate from and between any international airport on either side of the Tasman Sea. The operationalization of the agreement was postponed shortly before it was scheduled to occur by the Australian government in 1994 (Pearce 1995). Much of it was initialized later in 1996, however, with the exception of the rights of Australasian airlines to bring passengers into either country, fly them across the Tasman and out again to an external destination (Kissling 1998; Pearce 1995).

In terms of facility and infrastructure development, regional agreements should present some kind of positive effect. NAFTA should decrease the high tariffs on manufactured goods brought into Canada and Mexico from the United States that are used in tourism operations. It should also expand telecommunications links between member countries (G. Smith 1994). Reforms have already begun in ASEAN to liberalize air and sea transportation and telecommunications. Future plans include intra-ASEAN special fare deals, an increase in air links between secondary cities and tourism areas, improved regional infrastructure for cruise traffic, elimination of barriers that impede tourism growth (e.g. travel taxes), and proposed trans-ASEAN networks in rail, road, waterways, and telecommunications (ASEAN Secretariat 1999).

Taxes

Traditionally, one of the most obvious and bothersome issues facing trans-European travelers has been the wide range of value added tax (VAT) rates. VAT rates ranged from six percent in Dutch and Greek restaurants and hotels to 22 percent in Danish services, and the variation for food purchases were even wider (Mihalik 1992: 28). The European Commission is recommending that VAT rates be standardized throughout the entire Union. For some countries this will mean a lowering of tax rates, but for others it will mean an increase.

A single currency

While few of the world's economic alliances have established common currencies, this has long been a primary goal of the EU. During the past 15 years or so events have happened and treaties signed that have paved the way for establishment of a common European currency. In 1990 the Economic and Monetary Union (EMU) was created, and in 1994 the European Monetary Institute (EMI) was established to oversee the progress of the EMU. The euro was chosen in 1995 as the new currency unit among the 11 members of the EU who have elected to join the EMU (i.e. Austria, Belgium, Finland, France, Germany, Ireland, Italy, Luxembourg, the Netherlands, Portugal, and Spain), and the European Central Bank superceded the EMI in 1998 (Elliott 1997).

Since January 1, 1999, the euro has been operating as a recognized currency on world markets and has been widely used by commercial banks and other institutions for non-cash payments (European Central Bank 1999). On January 1, 2002, euro coins and banknotes will be put into circulation in the euro area, which includes the 11 states listed above who have adopted the euro as the Union's single currency.

According to the World Tourism Organization (1998), the switch to a single currency by most of the EU's member nations will have a positive impact on tourism by increasing competition and making prices more consistent throughout the region. The euro will allow tour operators, hotel chains, and car rental companies to operate in a broader market and with more open and even pricing. These conditions will create overall lower prices for tourism products and will thus potentially generate more tourism. Additionally, there will be the more practical advantage of not having to exchange currencies every time a border is crossed, which will save confusion and high exchange surcharges. According to one study, if a traveler were to begin a tour of all of the EU capitals except Dublin and Luxembourg with a sum of 40,000 Belgian francs, he/she would be required to pay no less than 18,800 francs in exchange costs alone (cited in Robinson and Mogendorff 1993: 23). Tourism businesses will also benefit for the same reasons, which should result in an eventual evening out of price ranges for tourism and other products throughout the Union (Hoseason 1998). According to Elliott (1997), the common currency is the crux of success for the European Union because many producers, including tour operators and other service providers, were reluctant to sell in other countries because they did not want to have to deal with foreign exchange hassles.

Environmental issues

While the environment is a visible component of most international alliances, it has not been well addressed in the context of tourism. The US–Canada FTA did not deal with the environment. Instead both countries have dealt with problems, such as acid rain in separate bilateral forums (Hufbauer and Schott 1992). NAFTA and its supplemental agreements, however, do a better job of addressing environmental concerns. Mutual cooperation and trilateral dialogue for conservation and the enforcement of environmental laws have been at the forefront of these negotiations, and all three parties realize the need to conserve the physical and social environments within which tourism develops (Henderson 1993; G. Smith 1994).

According to the ASEAN Secretariat (1999), ecological conservation will be enhanced within the Association because common approaches to address environmental management and protection will be designed among member nations in the tourism development process. Guidelines are expected to be drafted to assess and monitor the impacts of tourism on the cultural and natural environments. Along these lines, ASEAN goals include building public awareness through education programs in collaboration with media at the community level, and widening public–private partnerships to allow the involvement of local citizens in the planning process.

Promotion and marketing

International competitiveness is likely to increase when cross-boundary regions are permitted and encouraged to cooperate in areas of promotion and marketing. International economic alliances in effect form larger national market bases for travel within the union or abroad. Likewise, they become a larger destination, both spatially and demographically, which may bring in larger cohorts of tourists who are interested in staying longer and spending more (Timothy 1995a). As frontier barriers are brought down through multi-national arrangements, mergers and other forms of integration will probably occur and will assist in the creation of economies of scale, thereby boosting competitiveness and the resultant lowering of prices.

By linking the three countries of North America together, NAFTA is expected to promote global competitiveness for tourism among the three nations. Some observers believe that the larger North American market would help the industry be more competitive with international markets in Asia, Europe, and the rest of the world (Smith and Pizam 1998: 18). With a reduction in border procedures and the common aviation market, New Zealand and Australia have the potential to be promoted as a multi-destination area. New Zealand officials argue that the numbers of tourists will increase in the entire region, thereby having a positive effect on both countries as both are promoted as one large destination region (Pearce 1995). Similarly, one of the primary goals of the ASEAN Plan of Action for tourism is to create an image of ASEAN as one appealing destination that will provide travelers with a wide variety of experiences. Member states are beginning to bring down the borders towards a more complementary relationship in tourism marketing rather than the traditional competitive approach that has characterized the region for so long (Timothy 2000c). This will be achieved by promoting the whole region as a single destination, offering thematic tourism packages and attractions, as well as by other trans-regional activities (ASEAN Secretariat 1999).

The primary difficulty in creating large markets and destinations through international treaty is that each country must give up some degree of national interest in planning and marketing efforts for the good of the greater unit. Also, difficulties will arise pertaining to enumeration of visitors from the outside. As borders fall, it becomes increasingly difficult to count and monitor foreign visitation on a national level. Estimates become weak and data become available only on the level of the entire bloc. For many this will not be a problem, but for individual member countries, this information will essentially be unattainable.

The reality

While many international blocs provide legal mechanisms for major changes in the tourism industry, in most cases few effects can yet be felt on the ground, the European Union being the primary exception to this. In North America, for example, few objectives of the NAFTA agreement have had a major impact in practice and those that have been implemented are unnoticeable to the average traveler. Border procedures are still required by all tourists, and few prices have been lowered noticeably owing to increased investments, greater levels of competition, and a reduction in tariffs. Exchange rates, if anything, are more likely the cause of the most visible changes in travel costs in North America. The implementation of NAFTA so far has not effected a notable growth in tourist arrivals in member states. In fact, during the three years following the 1994 agreement, Mexican and Canadian arrivals to the United States saw an average annual decline of nearly two percent. Likewise, Mexican arrivals to Canada have stayed relatively even (Smith and Pizam 1998), except for extreme drops in 1995 nearly 25 percent, owing to the rapid devaluation of the Mexican peso that year. Tourism to Europe is continuing to grow, although it is difficult to know whether or not the implementation of EU policies has had any effect on this. Exchange rates have played an important role in this context as well, because with the strong US dollar, there has been a clear increase in North American trans-Atlantic travel in 2000 and the late 1990s.

While limited in scope, several binational partnerships have developed for the promotion of tourism, including the 'Two Nation Vacation' program between Texas (USA) and Nuevo Leon (Mexico) that focuses on their shared history and traditions. This, according to Smith and Pizam (1998) is one result of NAFTA.

Within the contexts of Europe, Southeast Asia, and North America, it is still too early to understand the effects of international economic and trade alliances on tourism. As Smith and Pizam (1998) observed, little exists in the academic literature on the effects so far of international alliances on tourism, and in the trade magazines, the anticipated effects are still being discussed.

Declining international relations

While the discussion above has pointed primarily at the positive side of change, it must be noted that not all geo-political change in recent years has been positive.

Wars

Chapter 2 discussed war as a barrier to tourism. Often relations between nations decline to such an extent that wars ensue and sometimes coexistent neighbors become bitter enemies. While relations between Canada and the United States are generally good, and relations have never soured to the extent of some other regions of the world, there are times when conflicts result in heated exchanges. For example, disputes over fishing rights between US and Canadian fishing fleets and customs officials have become so heated that they have resulted in gunfire exchange across the border at times (Vesilind 1990).

Animosity on both sides of Cyprus is intense and each side continues to blame the other for problems facing the island. Mounting mistrust between the two communities and propaganda on both sides continue to divide an island that used to be a thriving tourist destination and which still holds a great deal of potential for development in the future (Ioannides and Apostolopoulos 1999: 55). Even when relations between the two sides seem to be improving, events occur that thwart peace negotiations and threaten the existence of tourism in an already fragile political environment.

The wars associated with the dissolving of Yugoslavia obviously had an extremely strong impact on tourism to the region. Many of the most historic cities and individual attractions, such as Dubrovnik and the Mostar Bridge, were destroyed. Today, however, tourism is on the rise in nearly all areas of the former Yugoslavia (Hamilton and Broustas 1996), with the exception of Serbia, Kosovo, and Macedonia, which were all involved in conflict with NATO intervention (early 1999). Coastal regions have been the first areas to re-establish tourism, and many are beginning to thrive, while the interior is improving at a much slower pace. Once crowded with German and Austrian tourists heading for Greece, the main highways through the former Yugoslavia are now quiet. Several border crossings now require multiple visas for the trip, which, in addition to questionable safety, has effectively decreased the number of inland tourists (Hamilton and Broustas 1996).

Very telling is an account given by Gosar (1999) of his recent travels between Bosnia and Croatia as follows:

> In the summer of 1997 I decided to visit Dubrovnik, Sarajevo and the pilgrimage town of Medjugorje. The Adriatic coast is, in its long Croatian part, broken up with a 10 mile long stretch of land belonging to the Bosnian Federation. This was, within the former Yugoslavia, this Republic's window to the sea. The Bosnian township of Neum was developed into a middle-sized tourist resort. In accordance to the general agreement between former republics, the Official Bulletin of the Republic of Croatia even names two Croato-Bosnian border crossings of the second category on the highway Split-Dubrovnik ... As I moved towards Dubrovnik, I could not find any sign of an international border there. The only visible markings, where the border supposedly should exist, were two abandoned huts, a kind of military posts (sic), which nobody occupied. Later, as I turned inland towards the pilgrimage town of Medjugorje, nobody stopped me, despite the fact that this place, settled by Croats, is more than 15 kilometers within the border of the Federation of Bosnia and Herzegovina. We past (sic) Mostar, on the route to Sarajevo – more than 100 km inland – a strip and search

> police control was performed by both: the Croats of Bosnia and the Muslims of Bosnia.
>
> (Gosar 1999: 1-2)

Likewise, according to another traveler in the same region:

> I was trying to get from Sarajevo to Belgrade. You definitely cannot do it from the main bus or train station. But one way is to cross into the Republic of Srbska (still part of Bosnia). You can take a tram or taxi from downtown Sarajevo and cross on foot, passing the Republica Srbska control post, to Lukovica. Their you can take a bus ... I got hit, as we crossed from Bosnia Herzegovina into Yugoslavia, by the Republic of Srbska DEM 30,- 'transit visa' and, just a couple of minutes later, by DEM 62,- for a Yugoslav visa, which you can now obtain at the border. Of course, the helpful folks at the embassy in Bratislava, Slovakia told me US citizens don't require visas at all. But it could be due to a rapidly changing situation.
>
> (Gibson 1998: 3 quoted in Gosar 1999)

Increased barrier effects

While the decrease in barrier effects is improving generally throughout the world, in some cases things are regressing. Despite years of negotiations between the United States, Canada, and Mexico to establish the North American Free Trade Agreement (NAFTA), which is supposed to improve trade relations and allow for freer passage of citizens between member nations, the United States recently enacted a new law that has hurt US–Canadian relations (Griffith 1997). The new immigration law described in chapter 2 will require the documentation of all non-US citizens upon entry into the country, has become very unpopular among Canadians and American border residents. If implemented the law would result in grave economic problems between the two trading partners, and Canadian cross-border travel would be severely curtailed. US senators from the southwestern border states are eager to tighten border controls, while those along the US northern border are reluctant (*Travel and Tourism Executive Report* 1997).

Similarly, owing to Canada's economic connections to Cuba, the United States threatened to boycott its northern neighbor in 1998 by restricting access into the country from Canadian companies and business people who conduct business in Cuba. These two incidents have strained US–Canada relations in the past few years and increased the barrier effect of the 'longest undefended border in the world.'

Until 1996 several informal ports of entry existed along the US–Mexico border, where tourism thrived as Americans in particular crossed the river to spend the day in Mexico. A recent crackdown in illegal immigration and the war on drugs has led to these unofficial crossing points being off-limits (Herrick 1997). This has curbed the cross-border tourist trade and resulted in lost jobs and income levels in several Mexican border communities.

Similarly, in an era of peace making between Northern Ireland and the Republic, when the borders of the EU were supposed to be coming down, those in Ireland were being erected and reinforced. One newspaper commentary suggested that 'thirty years ago the Irish border was a joke and comedians wrote skits about their efforts to find this dividing line. Today it is a real barrier, growing ever stronger as Europe develops towards fulfillment of its slogan of a "Europe without frontier"' (quoted in O'Dowd *et al.* 1995: 282). This came about at a dangerous time in the clashes in Northern Ireland, when many of the rural roads crossing the border were being closed. These closures proved difficult for the areas in question that were dependent economically on agriculture and tourism (O'Dowd *et al.* 1995).

Landscape changes through political change

Just as the existence of political boundaries creates unique features and patterns in the human landscape, so do the changes discussed throughout this chapter.

Customs and immigration structures

In the EU, the changing role of international boundaries in recent years has created a landscape of abandonment on most borders in Western Europe. Structures that were once part of the landscape of functioning boundaries have become derelict evidence of political change and the development of the 'borderless' Union. As early as 1992, customs offices along the French–German border were closed. Only border police remained but rarely checked travelers' identification, and in 1993 all controls were supposed to disappear altogether (*Economist* 1992). In April 1995 the author traveled extensively across the borders of several countries within the EU and found that all customs and immigration offices within the Schengen countries had been abandoned, and many were in complete disrepair (Plate 5.1). Several frontier stations, however, had been converted into tourist information offices and small local museums. Thus a landscape reflecting years of jealous sovereign control and barriers to people, goods, and services, had become one of integration. In Berlin the same is true where, as mentioned earlier, Checkpoint Charlie, the notorious West Berlin border station, is now the centerpiece in the new Checkpoint Charlie Museum (Apple 1996; Kinzer 1994).

New border crossings are another sign of improved relations. Until 1994 only one official crossing point existed between Israel and Jordan but access was limited only to one-way traffic from Jordan into Israel. As a result of the

Plate 5.1 Many of the customs and immigration offices at border crossings in the EU have been abandoned, such as this customs agency on the Dutch–Belgian border (1995). In many locations, these structures have been transformed into museums and tourist information centers

1994 peace treaty, however, the two countries opened a new border crossing that for the first time allows third-country citizens to travel freely in both directions. The new crossing, Arava, is located about four kilometers north of Israel's Red Sea resort Eilat and Jordan's Aqaba. Jordan now accepts passports containing Israeli stamps, making travel possible in both directions (Kustanowitz 1994).

Methods of demarcation

While it is true that many militarized boundaries have become lines of peace and cooperation, such as the East–West German divide and the Austria–Hungary frontier, where images depicting Hungarian police dismantling barbed-wire fences separating the two countries dominated media headlines in 1989–90, many fortified boundaries have been erected and reinforced. Even though relations are generally good between the United States and Mexico, for instance, owing to perceived threats by illegal Mexican immigrants, the US Border Patrol recently refortified the border fence to include heavy steel backings along the former chain-link fence, concrete walls, and deep anti-vehicular trenches in some locations.

Obviously these types of changes do little to promote the peaceful image of tourism, but it is clear that the US government has something besides tourism on its agenda for that region.

Summary

Recent monumental changes throughout the world in geopolitical terms have brought about improved international relations in general and created circumstances where people nationalities that have long been denied entry to some countries (e.g. Americans to Albania) are now able to travel freely within former enemy states. In the same way, countries and regions of the world that were long off limits to foreigners (e.g. Tibet) are now opening up to a realize of their potential as new and 'undiscovered' tourist destinations.

As a result of the rapid process of globalization, dozens of economic and socio-political international alliances, or trading blocs, have been established (e.g. NAFTA, ASEAN, and EU). While most of these have economic issues as their primary objectives, which themselves have far-reaching implications for tourism, the policies they are establishing in the order of human resources, cross-frontier flows of tourists, environmental conservation, marketing, and transportation and infrastructure also directly impact the type of, and extent to which, tourism grows and matures.

6 Tourism planning in the borderlands

> We need to learn what is happening on the other side of the border. We need to learn what plans are contemplated and why, and then to sit down and see what we can do to reach some form of cooperation and mutually acceptable solutions ... Our planning world does not end at the border
>
> (Kjos 1986: 22, 26).

Introduction

Sound tourism planning is generally viewed as a way of mitigating the negative effects of tourism while at the same time enhancing the benefits. There has recently been a shift in traditional tourism planning paradigms and research from narrow concerns with physical planning and blind promotion aimed at the masses towards a more balanced approach that supports the development and promotion of more sustainable forms of tourism (Getz 1987; Inskeep 1991; Murphy 1985). In the context of tourism, sustainability can be viewed from two perspectives. First, the local physical, socio-cultural, and economic environments are treated in such a way that they will be maintained as viable resources and functioning systems in the long term. Second, the industry itself must be maintained at a healthy level into the future as well with minimal negative impacts.

Most tourism planning scholars agree that sustainable tourism development can best be accomplished by involving local residents in decision making and in the benefits of tourism, and by collaborating between various stakeholders in decision-making matters. Most characteristics of these participatory and collaborative planning approaches originated in the transactive and advocacy planning traditions, which herald the notion of giving residents and interest groups more control over their own destinies (Hudson 1979).

The academic literature argues that borderlands are unique situations where the theories and paradigms used to explain economic, political, social, and geographical patterns in most regions offer inadequate explanations. Planning too requires special consideration, and it is the cross-boundary cooperative and participatory approaches to tourism development that are most vital in frontier regions. This chapter examines the special tourism planning needs in border areas, reviews the benefits of cooperative and participatory planning, and summarizes the various challenges to both approaches.

The need for planning in border regions

Planning is important everywhere to enhance the positive aspects of development and mitigate the negatives. However, it is clear from the discussion so far that borderlands are highly complex, and evidence shows that they may merit special planning considerations. In Husbands' (1981) words, they are specialized problem areas. House (1980: 463) argued that borders are not just on the periphery, rather, a 'double peripherality' exists, where the effects of territorially marginal locations are compounded by the unfavorable conditions of a frontier location. Owing to their geographical location in their country's regional system, in most cases border zones have traditionally been regarded by national governments, industries, and residents as peripheral areas in socio-economic terms. This status, together with the political nature of boundaries, has created challenging conditions – many of which are distinct from those in other national spaces.

Owing to their barrier and filtering effect, frontiers create unusual economic opportunities and social conditions. For example, the number of immigrants in northern Mexico attracted to new jobs has caused a notable population boom, which has resulted in larger and more crowded squatter settlements in urban areas (Ingram *et al.* 1994). Also, the former peripheries are becoming areas of economic growth, based largely on private cross-border trade, services, and the growth of the bazaar economy. One result is that the actions of the informal sector in response to the new border situation are far ahead of the formal one (Stryjakiewicz 1998: 211).

Since borderlands are located on the purlieus of their respective national traffic and communications networks, they are frequently underdeveloped in the areas of transportation and infrastructure (Krätke 1998). Road networks, public transportation, and communications systems are often rudimentary in frontier regions, and national governments rarely make efforts to improve the conditions in these relatively 'unimportant' areas (Timothy and White 1999).

In some transfrontier urban areas residents are permitted to travel back and forth between sides for shopping, family contacts, work, and education. This has led to a need for international mass transit systems that can carry passengers from one side of a border to the other. However, cross-border transportation is difficult to coordinate. At the US–Mexico border, for example, conflicts have long ensued regarding the binational use of taxis and buses for public transportation. Concerns about insurance coverage have been a primary obstacle to this (Martinez 1988), although in 1990, after much negotiation, the El Paso-Juarez Trolley Company was established. The business is owned equally by investors from both countries, and it carries passengers across the border and throughout both cities along several daily routes (Maier 1992). Similarly, a trolley service between San Diego and the Mexican border city of Tijuana was also initiated recently and required significant cooperative efforts between the two communities (Sverdlik 1994).

According to Graizbord (1986), land uses should correspond with an area's inherent ecological attributes. Land that might be best used for agricultural production may be misallocated to some other use on the other side of the boundary, thereby causing ecological imbalances and managerial problems. He also makes the point that since land uses on one side of a border influence types of spatial development, and sometimes even land prices on the other, the environmental impacts of various land uses must be monitored and controlled.

Since human-created borders do not bind ecosystems, the use of natural resources requires joint management along an international border if this is to be done equitably and sustainably. Unfortunately, federal laws administered by the US Environmental Protection Agency tend to ignore unique border situations, and residents' interests in policy development are marginalized or ignored altogether (Ingram *et al.* 1994). Furthermore, according to Ingram *et al.* (1994), in environmental terms, borders commonly obstruct grassroots problem solving. They separate problems and solutions and aggravate perceived inequalities. Inequitable access to water resources is a particular concern on the US–Mexico border where the Colorado River and aquifers are being used up too quickly on the American side before Mexico gets a chance to use them.

Chapter 5 discussed recent geopolitical changes that have created massive movements towards globalization. This has allowed a greater level interaction between societies and given residents freedom to pursue various activities in contiguous regions. Thus, communities adjacent to international boundaries are becoming ever more integrated economically to the extent that cross-border regional economies are developing where trade and tourism exist nearly unhindered by the border (Artibise 1995; Herzog 1992; Kresl 1993). Herzog (1992: 16) maintains that 'while cities retain the elements of their nationally derived ecological structure in terms of density, social geography, road configurations, and physical design, they also display increasing patterns of connectivity across the border.' Generally speaking, cross-border employment and travel have never been simpler, and new institutions that facilitate transfrontier coordination are being created (e.g. trade blocs and Euroregions) (Stryjakiewicz 1998).

Because so many frontier cities are becoming more integrated, urban planning in borderlands is unique in the sense that it is necessary to resolve municipal-level problems through foreign policy channels (Herzog 1991c). For as

Hansen (1977) indicated, border regions cannot be accurately analyzed without viewing them as functional economic areas, the boundaries of which may not conform to official administrative boundaries. Nonetheless, differences in spatial structure of Mexican and American cities cause some concerns for planners and create issues that must be dealt with (Herzog 1991c).

Owing to these unique and somewhat disabling conditions, frontier zones, some experts argue (e.g. House 1980; Husbands 1981; Krätke 1998; Renard 1994; Sweedler 1994), require specific policies and actions from a planning and development standpoint. Herzog (1986a: 69) contends that 'planning for the border regions is crucial to the well-being of nations, in terms of both their economies and their public health, as well as to prevent political conflicts between nations.'

According to Husbands (1981), tourism is of particular importance in peripheral zones. It can be the key in spreading economic benefits to borderlands and integrating them into the mainstream national economy. The borderland periphery is 'just what modern tourism is looking for. This creates an (sic) unique chance for the marginal zones to catch up at least partly with the heartlands' (Bachvarov 1979: 135). Specifically, because tourism seeks out peripheral regions (Christaller 1963), special planning needs also exist in the context of tourism in border areas, to deal with issues like transportation and the relationships between tourism and existing economic activities (e.g. agriculture, forestry, and mining) (Ciaccio 1979).

As a result of the political changes described in the last chapter, many governments are beginning to pay more attention to border regions. These areas are much less likely today to be considered unimportant and peripheral in an economic sense. As the role of borders changes, they are becoming more lines of contact and thus zones of production, trade, and tourism. This is particularly the case in areas where frontiers have traditionally been closed and the communities on opposite sides isolated from each other. A good example is the Tumen River Delta, the common border of China, North Korea, and Russia. Industry, trade, and tourism have grown substantially in all three countries since the early 1990s, and are continuing to play an important role in the area's economic development (Sommers and Timothy 1999; Ting 1994).

It is clear that each of these issues has some implication for tourism, and it is argued here that nearly all of them can be addressed through proper planning. Cross-border cooperation and participatory planning are the two approaches to tourism development that best suit the situations described above.

Cooperation, collaboration, and cross-border planning

In regions where cultural and natural tourism resources lie across or adjacent to international boundaries, some form of cooperative planning is necessary if the goals and principles of sustainability (i.e. equity, efficiency, integration, balance, harmony, ecological integrity) are to be realized. Two sides of a border cannot ignore what transpires on the other side. In the words of one planner near the Mexico–US frontier

> Our planning world does not end at the border. We simply cannot afford to ignore what is going on to the south. Like it or not, we live in one region with shared problems and opportunities. If ignored long enough, problems ... will belong to everyone. The opportunities are likewise mutual, but if ignored long enough, then may be lost. We have much to do, but with growing awareness and growing understanding, we can and we will.
>
> (Kjos 1986: 26)

Cross-border partnerships take on a range of forms and occur at various scales. Institutionalized networks are endorsed or sponsored by government agencies that are authorized to act on an international level. These relationships are usually established bilaterally through legislative action. Informal cooperation, on the other hand, is established between local authorities, businesses, or individuals on two sides of a border, although it is not normally supported legally through official treaties and laws. The EU's Euroregions (e.g. the entire Germany–Poland border) are formal cross-border establishments whose purpose is to benefit borderlands in economic development terms (Bertram 1998; Koter 1994; Krätke 1998). Two communities on opposite sides of a border planning a joint sporting event or festival is an example of informal cooperation.

Martinez (1994) published a four-part typology of cross-frontier interaction in border regions. First, alienated borderlands exist where everyday communication and interaction are almost completely absent. Second, coexistent borderlands are those where the boundary is slightly open to negligible levels of interaction. Third, interdependent borders are characterized by willingness between contiguous countries to build cross-boundary partnerships. Fourth, integrated borderlands are defined as those where all major political and economic barriers have been abolished, resulting in the free flow of goods and people.

Timothy (1999a) takes this typology a step further by combining Martinez's (1994) alienation, coexistence, and integration elements with cooperation and collaboration to form a five-part model of levels of cross-border partnership in tourism (Figure 6.1). Alienation exists when little communication and no partnerships exist between neighbors. Cultural and/or political chasms are so wide that linkages are not possible or feasible. Coexistence involves minimal levels of partnership. Mutual toleration exists but nations do not interact harmoniously. Cooperative networks are characterized by initial efforts between adjacent administrations to solve mutual

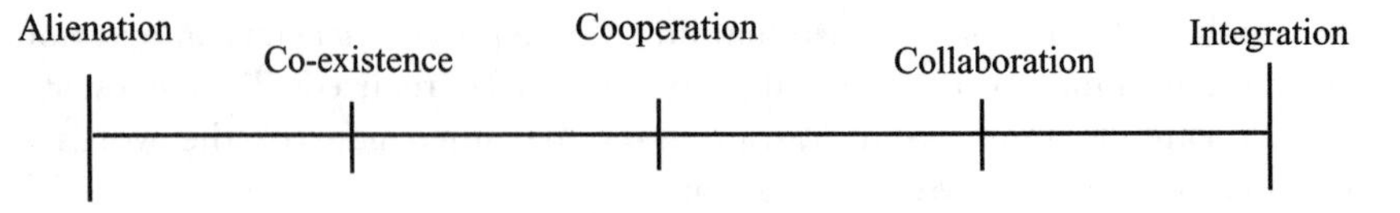

Figure 6.1 Levels of cross-border partnership
Source: Timothy 1999a

problems. Collaboration occurs in regions where relations are stable and joint efforts are well established. In these types of borderlands, partners work together on development issues and share some degree of equity in their relationship. Finally, integrated borderlands are those that exist with very few hindrances, and both sides are functionally coalesced. Each jurisdiction willingly waives some sovereignty in the name of common progress.

Although cooperation and collaboration in tourism are clearly important, not all outcomes are necessarily positive (Timothy 1999a). Cooperative efforts can be costly and time-consuming, and they sometimes result in effects not commensurate with the efforts involved. In some cases, cross-border partnerships can lead to political opportunism and the reinforcement of existing power among a privileged elite on one or both sides of the frontier (Church and Reid 1996: 1299). This can result in more pronounced disparities between areas in regional development terms (Scott 1998). Binational partnerships might also promote unhealthy competition that creates rivalries between local authorities, which can result in unstable cross-border relations (Church and Reid 1996; Krätke 1998). Additionally, Scott (1998: 620) argues that the formalization of cross-boundary partnerships may become complicated and lead to the bureaucratization of negotiation processes rather than permitting greater freedom to develop.

In spite of these disparagements, as mentioned earlier natural ecosystems and cultural areas are rarely confined within human-created boundaries. 'When such resources overlap international frontiers, yet another twist is added to the already complex concept of sustainability' (Timothy 1999a: 185). Thus, some level of interjurisdictional affiliation is crucial because it has the potential to diminish social, ecological, and economic disparities that exist on opposite sides of a boundary (Clement *et al.* 1999; Kiy and Wirth 1998) and will support more holistic and efficient planning as all parts of the attraction are considered one resource.

Achieving cooperation, collaboration, and integration, however, has been difficult in most borderlands because of the barrier effects of political frontiers. Thus, regional tourism tends to be competitive instead of complementary because it is basically self-contained in each country, and each side functions as a separate destination (Hussey 1991). The conventional view of competitiveness in economic development has meant that neighboring countries offer similar goods and services, so that they are in competition for the same or similar consumer base (Ho and So 1997; Kresl 1993; Perry 1991). Equally, complementarity refers to conditions where neighboring states or regions offer dissimilar services and products, creating a symbiotic, or complementary relationship.

Timothy (2001), however, argues that for some forms of borderlands tourism, this established economic view of trade and tourism is inadequate because borderlands are unique in many regards and the concepts are sometimes reversed. He suggests that two principle variables determine whether border tourism is complementary or competitive: the level of transfrontier partnership and the nature of the primary attraction.

With international parks and enclaves, complementarity exists because the attraction base is similar on both sides of the frontier – complementary because the two sides, working together, can create a local-level international destination. Competition exists when borders create discrete differences in political and social systems that spur activities such as gambling, prostitution, and shopping. These are competitive because communities on one side attract visitors across the border, while the other side tries to keep them from crossing. When one side finds success with an activity like gambling, the other side may create a similar activity in an effort to become competitive (Timothy 2001).

The second factor in determining the level of competition and complementarity is cross-border coordination. Public and private cross-border schemes have become commonplace (Church and Reid 1996; Leimgruber 1998; Scott 1993; 1998), and several benefits have been identified in the realm of tourism (Timothy 2000b).

Figure 6.2 illustrates that as the level of transfrontier networking increases along the alienation–integration continuum, so will the degree of complementarity to the extent that even the competitive activities described above could possibly become integrated, if both sides adopt them, or dissipated altogether if the activity is disallowed on both sides (Timothy 2001). Figure 6.2 also indicates that the journey from competition to complementarity is a dynamic process. Therefore, there is a possibility of regression towards alienation, as had happened recently along the Canada–US border where relations soured with the passing of the 1997 Illegal Immigration Reform and Immigrant Responsibility Act, which was discussed earlier in chapter 2.

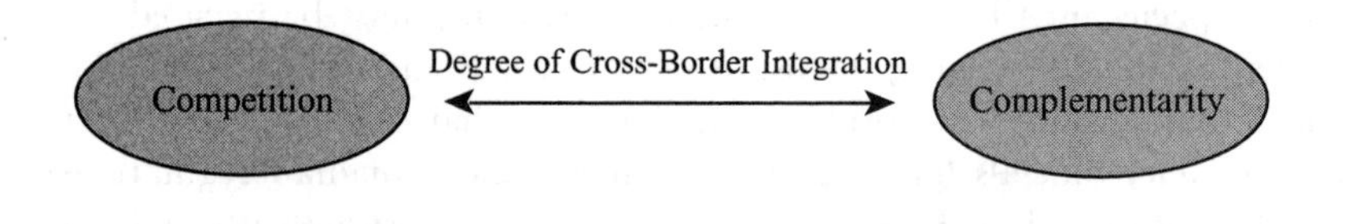

Figure 6.2 Competition–complementarity in the borderlands
Source: Timothy 2001

According to observers (e.g. Palomäki 1994; Tenhiälä 1994; Timothy 1999a; Wachowiak 1994), the following areas of partnership in tourism are particularly important in border regions because they are linked directly to borderland characteristics, contrasting political systems, and issues of sovereignty:

- Resource (natural and cultural) protection
- Infrastructure development
- Human resources
- Marketing and promotion
- Border restrictions and formalities.

Timothy (1999a; 2000b) provides an overview of the relationships between cross-border partnerships and sustainable tourism development. Some of the discussion in this section is taken from his work.

Natural and cultural resource protection

Nature and culture conservation can benefit significantly from cross-frontier partnerships. Binational collaboration can help standardize conservation regulations and controls on both sides of a border (Timothy 1999a) and promote ecological integrity, integration, balance, and harmony. This type of coordination will likely lead to more holistic and efficient planning as all parts of the ecosystem or attraction are considered one (Herzog 1992; Timothy 1998b). It will also help hinder the over-use of resources and eliminate to some degree the environmental, social, and economic inequities that exist on opposite sides of a border (Kiy and Wirth 1998; Timothy 1999a). This is particularly important when borders separate the location where problems are most obvious from the places where the most effective solutions can be applied (Ingram *et al.* 1994: 9).

Many examples demonstrate the growing importance of cross-boundary cooperation in resource management (e.g. Ellman and Robbins 1998; Herzog 1986a; Varady *et al.* 1996; Vogel 1997; Woodruffe 1998). In Waterton-Glacier International Peace Park, some natural features, such as Cameron Lake and Upper Waterton Lake, lie across the boundary – a situation that could potentially create conservation problems. For example, bull trout, a protected species in Alberta and if caught in Alberta, must be released. In Montana, on the other hand, the fish is not protected and can be kept. This has not created much of a problem in the lakes that straddle the border because Glacier officials have agreed to enforce the Canadian regulations on the American end of the lakes to avoid mixing up visitors with two sets of regulations (Lieff and Lusk 1990) and to promote the conservation of a common natural resource. Although nature conservancy regulations are somewhat dissimilar on either side of the border, administrators attempt to coordinate their efforts to work through the differences (Timothy 1999a).

In 1984 a remarkable event occurred as mayors from both sides of divided Nicosia, Cyprus, worked together to formulate the Nicosia Master Plan. The plan included goals to preserve the city's cultural heritage and to lure residents and businesses back to the decaying central city. It was an ambitious project, because the division of the city into two parts posed overwhelming political and technical challenges (Rossides 1995: 118). Nonetheless, the plan has been implemented, and both sides have so far been successful in adhering to it with a devout commitment to making it work (Kliot and Mansfeld 1997).

Infrastructure development

Transportation standards can be maintained better with the internationalization of infrastructure development (Artibise 1995; Wachowiak 1994). Open accessibility is obstructed when governments refuse to collaborate on issues such as road construction, transportation, and public services. In most cases, this is an effect of national priorities tending to outweigh cross-border needs (Naidu 1988). Networking in matters of infrastructure development can also decrease inequitable access to common resources (Ingram *et al.* 1994).

Efficiency would be improved through joint efforts as the costly and needless duplication of facilities and services, such as airports, hotels and shopping centers, were eliminated (Timothy 2000b). According to Anderson (1982: 13), international cooperation 'would permit a more efficient allocation of resources [and] prevent unnecessary duplication of facilities.' Gradus (1994) commented on this issue along the Jordan–Israel border where airports, highways, and utility services are replicated on opposite sides of the border, sometimes only meters apart. However, conditions are changing. The 1994 peace treaty between Jordan and Israel has fueled the growth of extensive cross-border cooperation. Dozens of joint endeavors between the two countries for the development of tourism have been initiated (Fineberg 1993; Government of Israel 1997; Gradus 1994; Kliot 1996; 1997). Recommendations include the expansion of the existing Aqaba airport (Jordan) into a common facility serving both countries' tourist resorts in the south, since there is more available land on the Jordanian side of the border. New terminals, a new runway, and a common customs facility are under consideration (Roberts 1995).

A similar situation already exists at the point where France, Germany, and Switzerland meet. The EuroAirport Basle-Mulhouse-Freiburg is located entirely on French territory but belongs to all three countries and services the entire region. Since Switzerland lies outside the European Union, the Swiss part of the airport is connected to Basle by a direct transit road that has no official border controls. Instead, Swiss immigration and customs procedures are conducted upon arrival at the airport before proceeding on to Switzerland (Walker 1994).

Human resources

Cross-frontier joint action in human resources can encourage more equitable and efficient management and improve ecological and cultural integrity as ideas are shared and knowledge gained through staff exchanges and shared training efforts. In addition, new jobs, increased regional income, and higher standards of living may result (Kaluski 1994). However, human resource partnership can be difficult to implement because frontier regions are sensitive areas in terms of illegal border crossings for employment, and protectionist restrictions keep potential laborers on their own side of the border (Timothy 1999a).

Human resource issues at the International Peace Garden are unique given the park's location on the US–Canada border. Staff members are permitted to work in either section of the park without having to obtain visas from the other country. This is possible owing to the park's position between the two countries' border stations. Thus, Americans are not required to pass through Canadian customs and immigration to get to work and vice versa. Each member of staff is paid and taxed based on her or his place of residence. For example, American staff are paid in US dollars and are taxed with US and North Dakota state income taxes. Workers' compensation, health, and unemployment insurances are also handled based on country of residence (Timothy 1999a).

In Waterton-Glacier International Peace Park, while not as integrated as the International Peace Garden, staff exchanges are common. Canadian interpretation personnel often spend their summers working in Glacier National Park (US side), while some American staff spend their summers working in Waterton. To avoid the need for work visas, workers are paid by their own park service while working abroad. These arrangements, it is believed, broaden the understanding of cross-national functions among personnel and contribute to establishing common goals and objectives (Timothy 1999a). Each year management from both sides meet together to discuss issues of common interest and to keep each other informed about events and managerial changes (Lieff and Lusk 1990).

Joint promotion and marketing

Efficiency, integration, and balance can result from cross-border collaboration in promotion and marketing. With the publication of joint promotional literature, budgets on both sides of a border can be rearranged so that the funds saved can be spent on other important obligations such as personnel, conservation, and infrastructure (Timothy 1999a; Wachowiak 1994). In addition, broadcast media spill across national frontiers, so joint efforts could possibly reach a wider audience and increase efficiency in promotional endeavors (Clark 1994).

Another perspective is the development of multinational destinations. Wachowiak (1994) contends that regional package holidays, festivals and cultural activities, and the development of cross-border attractions, such as trails and routes, are worthwhile foci of partnership. Ioannides and Apostolopoulos (1999: 56) argue that as global dissatisfaction with traditional destinations mounts, such as the Republic of Cyprus, new approaches to promotion and tourism development must be adopted. They recommend that in order for tourism on both sides of Cyprus to survive and flourish, for example, both sides need to work together to promote a single Cypriot tourism product, regardless of the political differences that have divided the island for so long.

These types of multination promotional efforts are important, particularly for destinations on the global or regional periphery, for as Tsang (1994, quoted in Heung 1997: 131) suggests regarding the pre-1997 alliance between Guangdong, Hong Kong, and Macau, 'this initiative has special significance for each of the destinations because the whole product would offer much more than the sum of the three areas each working on its own.' This type of collaborative marketing approach has the potential to maximize the tourism potential to the region and bring greater benefits to all parties involved in the long run (Heung 1997).

Most global examples of cross-border partnership in tourism have occurred within the arena of marketing and promotion. The ASEAN states have been exemplary in their efforts to promote the entire region as one large multination destination through international trade fairs, package tours, and promotional literature (Timothy 2000c), and the two sides of St Martin have done an excellent job of collaborating in the area of marketing (Office du Tourisme 1996; O'Neil 1996; *Travel Weekly* 1991). France and Belgium are working together to promote the Ardennes Mountain region of both countries. They have begun running tours on both sides of the border in an effort to cooperate, and brochures in each country laud the beauties of the landscapes on the other side of the border (Varniere-Simon 1991). Likewise, cooperation has recently strengthened between Northern Ireland and the Republic of Ireland. Several joint efforts are under way to promote the entire island as one destination (Cullen 1998), and a collective hotel guide for both parts of Ireland has been published (Milner and Brummer 1994).

In chapter 5 international economic communities and regional associations were examined. There is also a movement towards international tourism regions, whose primary aim is to promote and develop tourism on an international/regional level. The Baltic Sea Tourism Commission (BTC) is just one of hundreds of these already in existence. Begun originally in 1983, the organization included only the Baltic in its membership, but as interest in tourism to the region grew through the 1990s, and with the break-up of the Soviet Union, membership has grown to include 150 tour operators in 19 countries (BTC 1996). The primary aim of the BTC is to promote tourism to and within the Baltic Sea region, by engaging in regional promotional/marketing campaigns, developing cultural routes, and organizing special cruises and package tours. Marketing campaigns include regional representations at trade fairs, jointly published brochures, a website, and a multi-nation product manual.

A similar example is the Southern African Regional Tourism Council (SARTOC), with membership including South Africa, Lesotho, Swaziland, Malawi, and the Comoros. SARTOC's primary aim is to promote and coordinate the tourism industries of each member state. Special efforts are being made to encourage tourists from other continents to include most or all of the member countries in their itineraries while visiting southern Africa (South African Tourism Board 1999). Promotional campaigns include jointly-published literature and other forms of media that celebrate the beautiful landscapes, beaches, and cultures of each country (South African Tourism Board 1991).

Functional linkages in tourism terms between Mexican and American border towns have lately grown as an outgrowth of economic interdependence and environmental, social, cultural, and historic ties, and recent studies show that these linkages are growing stronger (Herzog 1992: 16). Promotional literature from San Diego strongly encourages visits to Tijuana and vice versa (Modler and Boisclair 1995). In one recent San Diego vacation booklet, four colorful pages extol the attractions of nearby Tijuana and North Baja California State, Mexico, providing information on attractions, activities, accommodations, and restaurants and encouraging tourists to visit in conjunction with a trip to San Diego (San Diego Convention and Visitors Bureau 1999). In this way, each city and region has become an added amenity for the other's tourism industries, and a form of interdependence has been created.

This interdependence is not new, for after the US–Mexican war of 1848, Tijuana was left on the Mexican side of the border, virtually inaccessible from the rest of Mexico. This created a south to north dependency relationship where Tijuana relied on San Diego for everything, including food, education, employment, and health care. This began Tijuana's long-time satellite relationship to San Diego (Martinez 1996; Puente 1996: 251). West and James (1983: 161) put it this way:

> Tourism's fortunes hinge, to a significant degree, on developments south of the border. For example, San Diego depends in part on attractions of Baja California, in general, and Tijuana, in particular, to lure tourists to San Diego. Tijuana, in turn, benefits and depends on attractions in San Diego to bring American tourists to it.

Recent years have seen a refocus in Mexican tourism development ventures towards the northern border. In 1990 Mexico initiated a new program to attract more visitors to its border towns and northern cities. Until that time the country's tourism officials had focused more on air travel to beach resorts, largely ignoring the borderlands, but now the situation has changed. Awareness-building campaigns are being launched to change people's perceptions of Mexican border towns, to develop a better and more inviting infrastructure, and to alter the lackadaisical, if inhospitable, attitudes of Mexican border officials (Budd 1990). To be successful, this program will require the combined nisus of communities on both sides of the border.

Border formalities and restrictions

As is discussed in the next section, border formalities and restrictions commonly deter the successful implementation of the types of partnership programs discussed in this section. Most of these areas of partnership can only be addressed adequately if some degree of lenience is given by the national governments involved. Border treaty waivers or less formal concessions will allow a more holistic form of planning and permit borderland tourism to function more efficiently and equitably (Timothy 1999a).

Several border restrictions have been eased in the International Peace Garden, without which, everyday functions would be nearly impossible. Canadian and US law requires a seven-meter wide swath, or vista, to be cleared of vegetation and built structures higher than one meter or so. This allows maintenance crews and surveyors easier access to the border markers. The International Boundary Commission (IBC) is a joint agency whose purpose is to maintain the border monuments and vista. The International Peace Garden successfully petitioned the IBC for permission to build a peace chapel directly on the borderline. The border marker for the IBC is located on top of the building so that the next marker down the line is in view. A similar waiver was recently granted when the Peace Park petitioned the IBC to allow a new welcome sign to be erected directly on the borderline (Timothy 1999a).

Other important, albeit somewhat trivial in appearance, treaty waivers have been sanctioned recently by the IBC. For instance, it is unlawful to transmit radio waves across the border intentionally. This created problems for Peace Garden employees who needed to speak by radio with work site supervisors or other workers who happened to be a few hundred meters away in the other country. In 1998 a waiver was extended, which allows park employees to communicate across the boundary within the park on radios for the purpose of handling everyday issues and business functions (Timothy 1999a).

Another important border concession in the International Peace Garden is the tax-free status of products to be used in the park. Because the Garden's gateway is located between the two customs depots, goods brought into the park from either country never pass through the other country's inspection procedures. Thus, all products are duty-free and can be used freely on either side of the border within park boundaries (Timothy 1999a). Similar frontier concessions have been made in Roosevelt Campobello International Park, also on the US–Canada border.

All of these areas of partnership have important implications for tourism. However, these are ideal situations. In most parts of the world cross-border cooperation, collaboration, and integration are lofty goals that are rarely achieved. The following section highlights some of the primary challenges facing cooperative planning in border regions.

Constraints to cross-border cooperation

Timothy (2000b) discussed several challenges facing cross-frontier collaboration within the context of international parks: political and cultural differences, border formalities and fortifications, sovereignty and territoriality, differing levels of development, and marginality of borderlands. These challenges, and others, are discussed below.

Political and cultural differences

Because values and cultures often differ on opposite sides of international boundaries, extra effort is required to bring to fruition cooperative efforts. Regardless of the suitability of a potential project for joint efforts, such projects still must operate in different cultural settings which are commonly marked on opposite sides of a border with differences in perceptions, values, and prejudices (Graizbord 1986: 14; Herzog, 1986b). Where symmetrical interests, cultures, and values exist on both sides, such efforts become much less burdensome (Blatter 1997; Scalapino 1992). Dissimilar economic structures, social systems, languages, and cultural traditions frequently hinder communications between sides. Sometimes these barriers are so great that partnerships become difficult to manage or are dissolved altogether (Krätke 1998; Saint-Germain 1995; Scott 1998). Political and social differences also result in contrasting planning regulations, building ordinances, labor laws, and pricing policies for raw materials (Wu 1998: 192), which can lead to conflict and a disintegration of cooperative efforts.

Administrative and organizational differences also can create difficult constraints to international collaboration. Achieving common goals in border regions is difficult when agencies in each country have contrasting mandates and opposing views of management and resource conservation. For instance, according to Weingrod (1994: 29), the US National Park Service and Forest Service attempt to preserve the natural environment while at the same time allowing recreational activities to take place. In Canada though, the agencies tend to be more polarized. Parks Canada functions much like the US agency, but the Forestry Ministry operates solely for the purpose of extracting resources.

Challenges also arise when different levels of government that are responsible for various aspects of planning and conservation meet at international frontiers (Timothy 2000b). In Mexico nature reserves and parks are usually controlled by states instead of the national government. In the United States, however, the federal government tends to control most preservation areas near the border. Likewise, the US borderlands are administered by state and federal agencies that do not always agree on border-related issues. The Bureau of Land Management, the National Parks Service, Customs, and Immigration, as well as the Mexican states of Sonora, Coahuila, and Chihuahua all have an interest in the everyday functions of the border (Steffens 1994). This creates a situation where 'a large number of individuals, groups, and agencies involved in all kinds of activities have an interest in the border region. Yet many of them seem to operate in almost total ignorance of the others. They duplicate each other's efforts, and their interests overlap' (Kjos 1986: 26).

Differences in interpretation of contractual meanings can further create communication barriers and lead to inaction or disputes between sides. For example, in environmental terms, international law requires that it is unlawful for one state to create transborder pollution that will cause serious damage in a neighboring state. However, the meaning of 'serious damage' may differ between states, particularly if the two nations are ideologically or developmentally different. 'If a downstream country is unwilling to tolerate levels of pollution which are deemed acceptable by the upstream country within its own borders does this constitute serous damage?' (Johnstone 1995: 52).

Border formalities

High tariffs and other border taxes are often established to provide a protectionist mechanism for domestic industries. This innately works against cooperation by creating barriers that make trade and tourism more difficult and costly. Travel restrictions on citizens of neighboring states also form a barrier to cooperation, particularly when formalities require burdensome inspections, and costly and tedious visa procedures.

Furthermore, border-related issues commonly take priority over other needs in frontier regions. National governments are generally more astute on preventing illegal migration and controlling the inflow of goods than they are on protecting the environment and establishing cross-border networks.

Sovereignty and territoriality

State boundaries and territory have been at the center of nearly every war fought throughout history, because countries are jealous about every fraction of national space, and sovereignty is commonly viewed as absolute state control of territory. Current and recent wars (e.g. Eritrea and Ethiopia, Iraq and Kuwait, Yugoslavia and NATO) started because of issues of sovereignty and territoriality. It is obvious, as discussed in chapter 2, that such altercations create barriers to tourism development. They also deter most types of international cooperation that might exist between states. Even in regions where cross-frontier linkages have existed for some time and where borders are well established, such as in the Rhine River basin, disputes over national sovereignty can inhibit cross-border partnerships (Scott 1989).

Border disputes and other forms of hostility, as discussed earlier, are significant barriers to tourism development. Clearly within the context of cross-border cooperation in planning, they also create formidable barriers, as in the recent conflicts along the Thailand–Myanmar border (Buszynski 1998), where there is a great deal of potential for tourism development and hence planning partnerships. Partnerships generally require the resolution of long-standing political conflicts and opposition. This may be very difficult to resolve and can thwart even the best and most supportive efforts for binational networking (Church and Reid 1995).

Truly integrated cross-national partnerships are difficult to achieve, even under friendly conditions, because both parties realize that they will be required to surrender some degree of sovereignty in the name of collaboration or integration (Blake 1994). Thus, there is generally no authority that can compel compliance (Johnstone 1995: 51). In one author's view

> By the beginning of the present decade, the political implications of the growing symbiosis between San Diego and Tijuana began to emerge more clearly. Although citizens, bureaucrats, scholars, and elected officials on both sides of the border recognized the need for local coordination, the principle of national sovereignty continued to impede the formation of any truly binational form of boundary governance. Thus, decisions would persist within the framework of separate jurisdictions of San Diego and Tijuana.
>
> (Herzog 1986b: 5)

In most parts of the world political traditions have dictated that all international negotiations are the rights and sole responsibilities of central governments (Gaines 1995), because sovereignty rights are owned only at the national level. This centralized view of power means that local authorities have few or no rights to formulate treaties or working agreements with their cross-frontier neighbors unless they have received prior authorization from the national government (Dupuy 1982; Hansen 1983; O'Dowd *et al.* 1995). On the island of St Martin there is almost a complete lack of coordinated efforts between the French and Dutch sides, with the exception of road maintenance. This results from the dependent status of each side of the island. Whatever is done jointly must be approved by French authorities in

Guadeloupe and Paris, and by Dutch authorities in Curacao and The Hague (Kersell *et al.* 1993: 51).

Even when decision-making powers exist on a local or regional level, they may not in the neighboring jurisdiction. Anderson (1982: 15) gave the example of France and Switzerland in the early 1980s, where communes, départements and regions in France did not have the legal authority to enter into any kind of cooperative agreements with their neighbors in Switzerland. The Swiss cantons, however, did have this authority, but lacking this right on the French side of the border made local level cross-border efforts unworkable. According to Gaines (1995) and Hansen (1983), this is unfortunate because the implementation of conservation and tourism policies and programs works best on a local level where mutual understanding and trust is stronger than at the national level, and residents tend to be more familiar with the local natural and cultural environments than bureaucrats from the distant and detached national capital (Timothy 2000b). Where they exist, strategies for accomplishing cross-border planning have tended to develop informally in local communities sharing a border (Briner 1991; Jessop 1995; Richard 1993; Sweedler 1994).

Differing levels of development

Contrasts on opposite sides of a border are clearest when developing country meets developed country. In most of the developed world environmental protection and pollution have long been a primary concern to citizen groups and governments, while the developing countries are still trying to find ways to feed and employ their people. Conservation thus becomes of secondary concern and is viewed as a luxury that cannot be afforded (Kjos 1986; Norton 1989).

Imbalances happen when one country has resources and know-how to devote to tourism development and management but its neighbor does not. In terms of parklands, Parent (1990: 33) suggests,

> Like Mexico itself, the Mexican park system is still developing. Mexico does not have the resources to staff and manage its parks as intensively as in the United States. Unlike the United States, Mexico cannot give such strong emphasis to environmental preservation. Instead, it must compromise more with economic development for the local people.

Differing levels of development sometimes correspond with varying environmental standards between neighbors (Timothy 2000b). Where pollution is uncontrolled on one side of a border, conservation efforts and tourism development are necessarily influenced on the other side (Steffens 1994). Along the US–Mexico border, for example, polluted water from Mexico threatens the health of residents on both sides (Ingram *et al.* 1994), and some recreational resources in the United States near the border are contaminated to the point where they are no longer useable for such purposes (Timothy 2000b).

The importance of the border in national economies also varies between developed and developing nations. The role of the frontier between the United States and Mexico is clearly an unbalanced one. It holds a much higher status in Mexico's national economy than it does in the US. Regardless of the fact that many American businesses are supported almost entirely by Mexican customers and that some institutions and agencies exist because of the border, the border from the American perspective is not a major player in the national economic system. In Mexico the opposite is true. The border is one of the fastest growing regions in the country in human and economic terms, and it has become the pride of the nation in many respects.

Marginality of borderlands

Frontier zones are generally viewed by national decision-makers as marginal and unimportant in their modernization and economic development efforts. An exception to this has been Mexico, where the frontier zone was the recipient of much development attention during 1960s and 1970s (Dillman 1970a; 1970b). However, typically throughout the world the more populated and industrial interiors are favored, which leads to a lack of administrative support and funding for economic development, including tourism in peripheral regions and borderlands (Blake 1993; Korona 1995). As mentioned earlier in this chapter, infrastructure, transportation, and communications systems are also commonly underdeveloped in border regions, which limits accessibility and blocks cross-border communications (Krätke 1998; Scott 1998), and bottlenecks frequently occur, creating crowded conditions and limiting the development potential of frontier areas (Wu 1998).

Peripherality also leads to the marginalization of border residents' concerns during policy development. Thus, it is not surprising that national level policies are often at odds with border needs and priorities (Ingram *et al.* 1994: 30), and the concerns and needs of border communities are ignored. Hansen (1977: 6) laments that what matters most 'is the narrow interests of the nation-state and not potential improvements in economic well-being that economic integration across national frontiers might bring to residents of border regions.'

Participatory tourism planning

This approach to planning should not be viewed in isolation from the partnership approach just described. Because border regions are often relegated to the economic, social, and political periphery, in addition to their physically peripheral locations, borderlands have traditionally been ignored in national planning. Thus frontier residents rarely have a voice in their own futures, particularly in the less developed world. Even in the developed world, residents from the national core tend to be better represented in decision making and policy development than are residents from rural, peripheral, and frontier districts.

Not only are national policies unresponsive to borderland needs (Ingram *et al.* 1994), they also inhibit bottom-up problem-solving efforts. The desire of local people to influence their living conditions is often unsupported and falls on deaf ears. Borderland residents are disadvantaged in essentially three ways: (1) they are usually distant from the capital or regional center; (2) their most favorable trading area is usually foreign rather than domestic; and (3) some of the local authorities with whom they need to deal are located in a foreign country.

Tourism in peripheral regions is often ultimately controlled, managed, and exploited by the developed industrial core, with little benefit for and control from within the frontier area (Keller 1987). While this center–periphery conflict is common, fortunately it is changing as more experience is gained with time, and as more governments are beginning to adopt more sustainable approaches to economic development (Timothy 1999c; Timothy and White 1999).

To eliminate, or at least to reduce, this effect it is important that the development process and decision-making authority remain firmly with the local authorities and people on the periphery. Tourism will be more sustainable, and hence more successful, if residents, local businesses, and other stakeholders are permitted to participate in tourism development (Bramwell and Sharman 1999; Gunn 1994; Inskeep 1991; Jamal and Getz 1995; Murphy 1985). Grassroots participation in tourism development can be viewed from at least two perspectives. First, residents must be empowered to determine their own goals, needs, and desires for development. Second, they must be given opportunities to benefit from tourism economically and socially (Brohman 1996; D'Amore 1983; Friedman 1992).

According to Long (1993), residents have to be involved in the process, otherwise they will alter even the most well-planned, well-meaning planning programs. Likewise, Gunn (1994: 111) suggests that 'plans will bear little fruit unless those most affected are involved from the start.' Murphy (1985: 153), one of the most outspoken advocates of the community approach to tourism, argues that 'tourism ... relies on the goodwill and cooperation of local people because they are part of its product. Where development and planning do not fit in with local aspirations and capacity, resistance and hostility can ... destroy the industry's potential altogether.' This is why Korten (1981: 613) argued that 'the more complex the problem, the greater the need for localized solutions and for value innovations – both of which call for broadly based participation in decision processes.'

It is very likely that if border residents were given more of a voice, many would choose to strengthen cross-border linkages, particularly if similar cultural groups were located on opposite sides of the border (Berg 1999). However, in some cases, this goes against national-level objectives.

Many examples exist where adjacent national governments have strained political relations, but on a local level, cross-border trade, travel, and other forms of informal partnerships exist (e.g. Greece and Turkey, Thailand and Myanmar, and China and Russia). In economic terms, borders usually suppress local residents from trading across the border with other nearby communities. Instead they are forced to trade inwardly, with the rest of the nation to which they belong, creating a fan-shaped interaction zone away from the border (House 1981). This can affect the quality and quantity of products available in the borderlands and determine the feasibility of tourism development. It may become difficult to acquire high-quality perishable food items (e.g. vegetables, milk, and meat), and the cost of goods and some services may be menacingly high owing to high transportation costs. Cross-border tourism clearly is affected in these situations, and in some cases this limits the region's global appeal to the extent that domestic tourists are the primary market base for the region (e.g. the former Soviet Union). Whereas, local-level cross-border trade and cooperation might eliminate some of these problems by allowing freer accessibility to other markets.

Constraints to participatory planning in frontier regions

When developed nation meets developing nation, or when two ideological systems (e.g. communism and capitalism) face each other, planning traditions are so different that difficulties arise. According to Herzog (1982: 863), the coordination of planning between the US and Mexico is complex owing to political traditions in decision making and power structure. Tijuana's planning issues are controlled by federal and state agencies rather than the local region, while in San Diego, the opposite is true. Decisions are made more often by local citizens, and public officials give more grassroots control over planning decisions to community members.

A lack of expertise on the part of local officials is a major constraint to allowing participatory planning. Many officials in developing and peripheral regions are unaware of the need to involve the public in decision making, and many are ignorant about how to do this. Sometimes, grassroots participation is misinterpreted solely to include educated researchers, trained consultants, and government officials (Timothy 1999c). This inclination supports Graf's (1992) claim that modernization efforts that underlie many government plans focus on the elites. No doubt the relative infancy of tourism in some borderlands and peripheral regions has led to inadequate expertise in these areas.

The power structure is also a primary concern in participatory planning. Many societies are in many ways locked into traditions that do not easily allow participation to occur. On the island of Java, for example, one of the most powerful traditions is that of authority and reverence towards people in positions of power (although recent upheavals in 1999 and 2000 appear to be calling this tradition into question). Javanese culture accepts that power and authority lie in the hands of the center (Liddle and Mallarangeng 1997; Moedjanto 1986). For there is

> no inherent contradiction between the accumulation of central power and the well being of the collectivity, indeed the two are interrelated. The welfare of the collectivity does not depend on the activities of its individual components but on the concentrated energy of the center. The center's fundamental obligation is to itself. If this obligation is fulfilled, popular welfare will necessarily be assured.
>
> (Anderson 1972: 52)

This proverbial approach to power appears to be common in most developing countries (Brown 1994; Kamrava 1993). According to de Kadt (1979), the experiences of local residents in political decision making for tourism in most countries of the developing world tend to be limited, owing to dominant national and local groups that purposely keep them in a subordinate position. Haywood (1988) agrees with de Kadt and argues that public participation is significantly hampered in many countries where officials have little interest in encouraging representational democracy. As a result, most people remain politically unprepared for involvement in the decision-making process (Timothy 1999c). The same could be said of border and other peripheral regions where grassroots participation is limited and, as a result, where the people are largely unprepared to be included in decision making (Hansen 1986; Høivik and Heiberg 1980; Keller 1984; Scott 1998).

Summary

This chapter clearly describes why border regions need special considerations from a planning perspective. Owing to their location on the national periphery in physical terms and on the national fringe in socio-economic terms, frontier regions are often ignored by central governments. This is evident in the typically inadequate infrastructure and transportation development, land use difficulties, and environmental problems.

Many of these challenges can be overcome through cross-border partnerships and participatory planning, where all stakeholders are involved. Success in cross-national planning is critical to minimizing the imbalances created by the overuse and pollution of natural resources on one side of the border.

Similarly, land-use conflicts can be solved and infrastructure problems addressed when administrations from both sides of international frontiers engage in collaborative efforts. Principles of sustainability (e.g. equity, balance, harmony, holistic planning, and cultural and environmental conservation) will generally be espoused and encouraged when cross-border cooperation is put into practice and when residents, who, owing to their customarily marginal role in national functions have not been involved in planning, are given a voice in decision making.

7 Conclusion

The future

> While you can't book a lunar getaway today, the prospect of vacationing on the moon isn't a flight of imagination either
>
> (Wilson 2000: 97)

There should be little doubt that borders are complex and influence tourism in a variety of dynamic ways: they are barriers to tourism, tourist attractions, and modifiers of the tourism landscape. As described in chapter 2, the barrier effects of international boundaries are unmistakable, hindering both the flow of tourists and the physical and socio-economic development of tourism in destination regions. While many world frontiers are difficult to cross owing to defensive demarcation methods and strict control measures, even the friendliest of borders can create psychological barriers for many people.

According to chapter 3, some of the world's most extraordinary tourist attractions lie either directly on, or in close proximity to, international boundaries. Niagara Falls and Maasai Mara-Serengeti International Park are prime examples of this. Likewise, certain tourist-oriented activities nearly always develop (e.g. shopping, prostitution, gambling, and drinking) adjacent to political lines when laws and policies pertaining to them are different on opposite sides, and when people are permitted to cross unhindered. On even a more specific level, the borderline itself can be an important tourist attraction because it presents some kind of curiosity in the cultural landscape and connotes differences in political systems, social mores, cultural traditions, and possibly ecosystems. Borderlands therefore hold a great deal of potential for tourism development, although on a global scale little has been accomplished in this arena.

The landscapes of tourism were the focus of chapter 4. These landscapes are very often distinct on opposite sides of a border based on differences in tenure systems, planning policies and traditions, settlement patterns, urban structures, and levels of socio-economic development. Often the barrier effect of borders is so great that parallel tourism development occurs where like services, infrastructure, and even attractions exist side by side but on opposite faces of a political divide with little cross-border coordinated efforts to link the two systems. In addition to borders influencing the tourism landscape, tourism in some cases has been crucial in effecting changes to the border landscape and its functions. This reverse relationship is likely to continue in the future as tourism continues to grow and play a more important role in bringing down political barriers that have existed for centuries.

These relationships (i.e. barriers, attractions, and landscape modifiers) are dynamic, and the current global economic and political climate has a major role to play in this fluidity. The focus of chapter 5 was the contemporary geopolitical changes that affect the growth and development of global tourism. Improvements in international relations and the collapse of communism in Eastern Europe have increased levels of freedom for millions of people who were previously home-country bound and for people of nationalities from which permission to enter has long been denied. Likewise, the creation of supranational alliances (e.g. NAFTA, ASEAN, EU) has led to more cross-border cooperation, more liberal travel and development policies, and more consistent levels of environmental and safety standards globally. Evidence strongly suggests that all of these events and actions profoundly affect the flow of tourists and the development of tourism in destination regions. This is especially clear when considering the abrupt and rapid increase in tourist numbers to and from Eastern Europe in the 1990s and the development of tourist communities in places like Albania that have long been off limits to mass tourism.

The unique planning needs of border areas were the focus of chapter 6. While planning is important in all regions, borderlands have special needs that must be considered in tourism development. Cross-frontier cooperation is particularly consequential in assuring that the principles of sustainability (e.g. equity, harmony, holistic development, and ecological and cultural integrity) are supported. Likewise, participatory planning is vital, particularly since borderland residents and environments have traditionally been viewed as marginal to national interests.

Future directions

This book has attempted to provide an overview of most of the concepts and issues that exist in border regions, and to highlight the primary relationships between political boundaries and tourism. As this text demonstrates, the idea of tourism and political boundaries is rich in concepts and theories, but there is a general lack of data for examining borders as destinations, barriers, and modifiers of the tourism landscape. It is hoped that this work will stimulate more scholarly thought about this inimitable yet commonplace situation, for the more information that can be collected and interpreted, the more we will understand the global nature of tourism.

While the work has focused primarily on the traditional view of political boundaries, it is important to remember that other types of frontiers exist that have not been included within the scope of this discussion. For example, cultural and gender boundaries are highly political and dynamic, and thus from a tourism perspective deserve additional research attention.

Boundaries that have long existed in defining womanhood and femininity in both the developed and developing worlds are today becoming much less of a socio-political barrier to opportunities in the production and consumption of tourism (Apostolopoulos *et al.* 2001). More women are now involved in tourism production than ever before, although in less developed countries, informal sector activities characterize most women's work in the industry. While this has improved the economic lot of many women, it has also increased their workloads dramatically for they are still responsible for their domestic chores as well. Crossing borders (i.e. the consumption of tourism), which has customarily been a male-dominated event, is becoming less gendered as more and more women are having opportunities to travel within their own countries and abroad. Greater numbers of women are attaining higher levels of education, are being placed in upper ranks of management, and are better off financially. The relationships between gender boundaries and tourism, particularly in the instance of women as consumers of tourism, have not been well researched commensurate with their socio-political prominence.

Cultural frontiers within countries and regions may exert just as much of a barrier effect as traditional political boundaries. The linguistic border between Anglo and Franco Canadians, for example, which does not necessarily correspond to established political lines, is a serious chasm between many people of the same nationality (Cartwright 1988; 1996) and has formed much of the foundation of Quebec's pro-independence movement. Similar situations exist in Belgium and other European and Asian countries. Correspondingly, religious divisions are typically blamed for the violent exchanges and political problems that have plagued Northern Ireland for many years (Boal 1994; Douglas 1998; Livingston *et al.* 1998). Linguistic and religious lines commonly determine the political attitudes, societal values, and belief systems ingrained in people's everyday experiences. Certainly these divisions are important within the realm of tourism, for they affect the host–guest relationship and create images (positive and negative) that have long-lasting consequences.

Culture as tourist attraction and the cultural impacts of tourism have recently received considerable attention in the literature. Part of the intrigue associated with visiting cultures that are different from the tourists' must lie within the concept of crossing ethnic boundaries, for most tourists travel in search of the different, the 'other', and the exotic – something beyond their everyday experience (Olsen and Timothy 1999). Thus, cultural boundaries form the basis of much tourist activity today. For instance, the Amish lifestyle, which is found in various places in North America and characterized by unique agricultural practices, transportation modes, styles of dress, language, and food, has become a highly important ethnic tourist attraction in parts of Pennsylvania and Ohio (Brandt and Gallagher 1993; Hovinen 1995). For Anglo Americans, a visit to 'Pennsylvania Dutch' country is like stepping into the past, a foreign past, where symbolic frontiers are crossed but where visitors can still drive in miles, spend US dollars, eat American fast food, and get along speaking their own version of English. Our understanding of the myriad relationships between tourism and language, religion and other elements of culture is in its infancy. There is a great deal of work to be done along this genre of boundaries.

A trend identified in this book, supranationalism, will no doubt become central in upcoming discourses on the globalization of tourism. As existing international alliances are strengthened and new ones created, scholars will have to be more cognizant of the effects of this on the industry. Labor migration, environmental management, education, and economic activities such as trade in goods and services, which all have primary functions within the production and consumption of tourism, will be significantly affected as so-called 'borderless' regions become more commonplace. So far, however, with the exception of the European Union, few international trading blocs have been successful in reaching high levels of integration (Timothy 2000c). Several international trade organizations have interests in tourism, including the International Monetary Fund (IMF), the General Agreement on Tariffs and Trade (GATT), and the Organisation for Economic Cooperation and Development (OECD). In fact, the Uruguay round of the GATT in 1993 paid special attention to mechanisms to encourage freer trade in services with important implications for tourism (Hall 1994b: 63). It is likely that tourism will come closer to the forefront of these multilateral negotiations as they continue to develop and as their mandates continue to be implemented, creating a rich subject area for additional inquiry.

Tourists are now beginning to infiltrate the last remaining 'frontiers' of the world. The industry in the high arctic and Antarctica, for example, is booming. With the exception of sub-oceanic regions, outer space is essentially tourism's 'final frontier' – a condition that may not last long, according to the quote at the outset of this chapter. However unrealistic it may appear at present, several hotel and resort companies have already begun seriously considering space travel. Extraterrestrial resorts, according to some analysts, are not as farfetched as critics might argue (Smith 2000). A 1997 NASA study concluded that interplanetary travelers represent a potential market worth billions of dollars, once economic and technical barriers have been overcome (White 1999). Companies like Hilton and Budget Suites have already begun pondering such options. One California enterprise has proposed using empty space shuttle fuel tanks to build an orbiting hotel (Wilson 2000). While skeptics reason that boredom will set in after the first day or so, the company argues that plenty of activities will be available to keep tourists entertained. Guests will be able to take space walks, study outer space, try their hand at space gardening, gaze down at the earth, and lend a hand in minor maintenance projects (White 1999: 10). While these prospects seem unlikely in the very near future, some estimates suggest that the frontiers of outer space might become viable destinations within the next 15 to 20 years (Wilson 2000). It will be interesting to monitor what the future holds for interplanetary travel. As the world becomes more affluent, as people become more demanding of hard-core adventure, and as crossing even the remotest terrestrial frontiers becomes more commonplace, society will continue to push its limits in search of new peripheries looking for new boundaries to cross and places to collect.

Finally, there is the issue of virtual travel, which implies the creation of new social spaces in the contemporary world and which allows a person to experience computer-generated places without leaving home (Rojek 1998: 40). Home entertainment technology is so advanced that people can now, or will soon be able to, cross borders virtually anywhere in the world. People can thus spend a week-long winter holiday on the beaches of Jamaica in only one afternoon without ever having to leave the comfort of their own living rooms. While this will unlikely completely replace the 'real' vacation experience, it may for some people become a more affordable and less stressful alternative. In this sense, virtual reality, according to Williams and Shaw (2000: 239), 'challenges the meaning not only of place but of tourism itself.' Thus, cyber-boundaries can now be crossed, and with the adoption of more advanced virtual technology it will no doubt become more commonplace.

This book is about the human experience in a socio-political world. Boundaries are human creations and thus a product of experience. From outer space, they are meaningless and invisible, yet in the terrestrial world they are replete with meaning and influence. The book is also about tourism. Its goal is to provide a step towards understanding these relationships. Borders are dynamic – they are in a state of constant flux. This suggests that some of the cases described here will eventually change, but the concepts will surely remain constant, for borders will continue to be barriers to, and venues for, tourist experiences as long as they continue to divide peoples and places.

References

Adams, K.D. (1995) 'Interstate gambling: Can states stop the run for the border?', *Emory Law Journal,* 44(3): 1025–67.

Agafonov, S. (1996) 'Border law: Pay before you cross', *Izvestia,* 10 December: 1–2.

Ahmed, Z.U. and Corrigan, F. (1995) 'An international marketing perspective of Canadian tourists' shopping behaviour: Minot (North Dakota) a case in point', in Z.U. Ahmed (ed.) *The Business of International Tourism,* Minot, ND: Institute for International Business, Minot State University, 96–123.

Ahmed, Z.U., Heller, V.L. and Hughes, K.A. (1999) 'South Africa's hotel industry: Opportunities and challenges for international companies', *Cornell Hotel and Restaurant Administration Quarterly,* 40(1): 74–85.

Airy, D. and Shackley, M. (1997) 'Tourism development in Uzbekistan', *Tourism Management,* 18: 199–208.

Akis, S. and Warner, J. (1994) 'A descriptive analysis of North Cyprus tourism', *Tourism Management,* 15: 379–88.

Alaska Department of Commerce and Economic Develoment (1989) *Alaska, 1990 Official State Vacation Planner,* Juneau: Department of Commerce and Economic Development.

Alberta Business (1992) 'Cross border shopping in reverse', *Alberta Business,* 9(1): 8–10.

Allcock, J.B., Arnold, G., Day, A.J., Lewis, D.S., Poultney, L., Rance, R., and Sagar, D.J. (1992) *Border and Territorial Disputes,* Third Edition, London: Longman.

Anderson, B.R.O. (1972) 'The idea of power in Javanese culture', in C. Holt (ed.) *Culture and Politics in Indonesia,* Ithaca, NY: Cornell University Press, 1–70.

Anderson, J. (1994) 'The great southwestern loop', *Travel & Leisure,* 24(8): 68–108.

Anderson, M. (1982) 'The political problems of frontier regions', *West European Politics,* 5: 1–17.

Andrew, B.H. (1949) 'Some queries concerning the Texas–Louisiana Sabine boundary', *Southwestern Historical Quarterly,* 53: 1–18.

Andronicou, A. (1979) 'Tourism in Cyprus', in E. de Kadt (ed.) *Tourism: Passport to Development?,* London: Oxford University Press, 237–64.

Ante, U. (1982) 'Der Beitrag der politischen Geographie zur untersuchung von Grenzräumen mit Beispielen aus Nordbayern', in H. Leser and H. Oeggerli (eds) *Deutscher Schulgeographentag Basel – Lörrach,* Basel: Geographisches Institut der Universität Basel, 59–66.

Apostolopoulos, Y., Sönmez, S. and Timothy, D.J. (eds) (2001) *Women as Producers and Consumers of Tourism in Developing Regions,* Westport, CT: Praeger.

Apple, R.W. (1996) 'Berlin: A fair city emerges out of the century's conflicts', *Travel Holiday*, 179(3): 58–9.

Appleby, T. (1995) 'Bordering on a reversal of fortune', *The Globe and Mail* (Toronto), 10 November.

Arnould, E. and Perrin, S. (1993) 'Developpement touristique et dimension transfrontaliere: le cas de l'espace Gaume-Meuse du Nord', *Revue Geographique de l'Est*, 33(3): 191–204.

Aronson, M.L. (1971) *How to Overcome Your Fear of Flying*, New York: Hawthorn Books.

Arreola, D.D. and Curtis, J.R. (1993) *The Mexican Border Cities: Landscape Anatomy and Place Personality*, Tucson: University of Arizona Press.

Arreola, D.D. and Madsen, K. (1999) 'Variability of tourist attractiveness along an international boundary: Sonora, Mexico border towns', *Visions in Leisure and Business*, 17(4): 19–31.

Artibise, L.F.J. (1995) 'Achieving sustainability in Cascadia: An emerging model of urban growth management in the Vancouver–Seattle–Portland corridor', in P.K. Kresl and G. Gappert (eds) *North American Cities and the Global Economy: Challenges and Opportunities*, Thousand Oaks, CA: Sage Publications, 221–50.

Ascher, B. (1984) 'Obstacles to international travel and tourism', *Journal of Travel Research*, 22(1): 2–16.

ASEAN Secretariat (1999 – updated) *Economic Cooperation*. Accessed 24 November 1999. http://www.aseansec.org/economic/eco.htm

Asgary, N., de Los Santos, G., Vincent, V., and Davila, V. (1997) 'The determinants of expenditures by Mexican visitors to the border cities of Texas', *Tourism Economics*, 3(4): 319–28.

Ashworth, G.J. (1995) 'Heritage, tourism and Europe: A European future for a European past?', in D.T. Herbert (ed.) *Heritage, Tourism and Society*, London: Mansell, 68–84.

Associated Press (1998) 'Drop in Canadian dollar draws U.S. shoppers across border', *The Toledo Blade*, 3 September.

Austin, J.P. (1979) 'Laredo: Trade center on the border', *Texas Business Review*, 53(2): 57–60.

Azienda Turistica (1994) *Campione d'Italia: Destined To Be Different*, Campione: Azienda Turistica.

Baarle-Nassau Tourist Office (n.d.) *Baarle-Nassau-Hertog: A Remarkable Village in an Enchanting District*, Baarle-Nassau: Tourist Office.

Bachvarov, M. (1979) 'The tourist traffic between the Balkan States and the role of the frontiers', in G. Gruber, H. Lamping, W. Lutz, J. Matznetter, and K. Vorlaufer (eds) *Tourism and Borders: Proceedings of the Meeting of the IGU Working Group - Geography of Tourism and Recreation*, Frankfurt: Institut für Wirtschafts- und Sozialgeographie der Johann Wolfgang Goethe Universität, 129–40.

Bachvarov, M. (1997) 'End of the model? Tourism in post-communist Bulgaria', *Tourism Management*, 18: 43–50.

Baerresen, D.W. (1983) 'The economy', in E.R. Stoddard, R.L. Nostrand, and J.P. West (eds) *Borderlands Sourcebook: A Guide to the Literature on Northern Mexico and the American Southwest*, Norman: University of Oklahoma Press, 121–4.

Bailey, R. (1995) 'A new air transport agreement between the United States and Canada has made it easier to fly over the border', *Air Line Pilot*, 64(7): 24–6.

Baldacchino, G. (1993) 'Bursting the bubble: The pseudo-development strategies of microstates', *Development and Change*, 24(1): 29–51.

Baldacchino, G. (1994) 'Peculiar human resource management practices? A case study of a microstate hotel', *Tourism Management*, 15: 46–52.

Bank of Canada (1985–98) *Quarterly Report*, Ottawa: Bank of Canada.

Barker, D. and Miller, D.J. (1995) 'Farming on the fringe: Small-scale agriculture on the edge of the Cockpit Country', in D. Barker and D.F.M. McGregor (eds) *Environment and Development in the Caribbean: Geographical Perspectives*, Kingston: University of the West Indies Press, 271–92.

Bar-On, R. (1988) 'International day trip, including cruise passenger excursions', *Revue de Tourisme*, 43(4): 12–17.

Barr, C.W. (1997) 'Old foes Vietnam and China cautiously rebuild ties', *Christian Science Monitor*, 3 October: 7.

Battisti, G. (1979) 'Tourism on the border between Italy and Yugoslavia: Methods for research', in G. Gruber, H. Lamping, W. Lutz, J. Matznetter, and K. Vorlaufer (eds) *Tourism and Borders: Proceedings of the Meeting of the IGU Working Group – Geography of Tourism and Recreation*, Frankfurt: Institut für Wirtschafts- und Sozialgeographie der Johann Wolfgang Goethe Universität, 215–29.

Baum, T. (1997) 'The fascination of islands: A tourist perspective', in D.G. Lockhart and D. Drakakis-Smith (eds) *Island Tourism: Trends and Prospects*, London: Pinter, 21–35.

Beers, C. (1995) 'Gold crown?', *Wyoming Wildlife*, 59(10): 10–19.

Beijing Review (1994) 'Tibet opens further to the outside world', *Beijing Review*, 37(40): 31–2.

Belden, T. (1997) 'European Union looks into details of airline alliances', *Hartford Courant*, 27 January.

Beltrame, J. (1997) 'U.S. delays border-control law', *Kitchener-Waterloo Record*, 10 November.

Belyy, N. (1996) 'Cross-border trade with China', *Rossiyskiye Vesti*, 18 April: 3.

Berg, E. (1999) 'National interests and local needs in a divided Setumaa: Behind the narratives', in H. Eskelinen, I. Liikanen, and J. Oksa (eds) *Curtains of Iron and Gold: Reconstructing Borders and Scales of Interaction*, Aldershot: Avebury, 167–77.

Bertram, H. (1998) 'Double transformation at the eastern border of the EU: The case of the Euroregion pro Europa Viadrina', *GeoJournal*, 44(3): 215–24.

Biddle, F.M. (1991) 'With "border war" over, councils from N.H., Mass. meets on tourism', *Boston Globe*, 7 October: 14.

Billington, M. (1959) 'The Red River boundary controversy', *Southwestern Historical Quarterly*, 62: 356–63.

Bixby, L. (1996) 'A new player at the table, the Sun Casino: Foxwoods gets a rival', *Hartford Courant*, 5 October: 1, 10–11.

Bixby, L. (1997) 'Foxwoods reports slot revenue increase for January', *Hartford Courant*, 4 February: 2.

Blacksell, M. (1998) 'Redrawing the political map', in D. Pinder (ed.) *The New Europe: Economy, Society and Environment*, Chichester: Wiley, 23–42.

Blake, G. (1993) 'Transfrontier collaboration: A worldwide survey', in A.H. Westing (ed.) *Transfrontier Reserves for Peace and Nature: A Contribution to Human Security*, Nairobi: United Nations Environment Programme, 35–48.

Blake, G. (1994) 'International transboundary collaborative ventures', in W.A. Gallusser (ed.) *Political Boundaries and Coexistence*, Bern: Peter Lang, 359–71.

Blatter, J. (1997) 'Explaining crossborder cooperation: A border-focused and border-external approach', *Journal of Borderlands Studies*, 7(1/2): 151–74.

Boal, F.W. (1994) 'Encapsulation: Urban dimensions of national conflict', in S. Dunn (ed.) *Managing Divided Cities*, Keele: Ryburn, 30–40.

Boggs, S.W. (1932) 'Boundary functions and the principles of boundary making', *Annals of the Association of American Geographers*, 22(1): 48–9.

Boggs, S.W. (1940) *International Boundaries: A Study of Boundary Functions and Problems*, New York: Columbia University Press.

Boisvert, M. and Thirsk, W. (1994) 'Border taxes, cross-border shopping, and the differential incidence of the GST', *Canadian Tax Journal*, 42(5): 1276–93.

Bondi, N. (1997) 'Shoppers head to Windsor for deals: High exchange rate on American dollar fuels big-ticket buys', *Detroit News*, 30 December.

Bondi, N. (1998) 'Bargain hunters hit Canada: Loonie's record fall an economic windfall for Americans shopping across the border', *Detroit News*, 30 January.

Borneman, J. (1998) 'Grenzregime (border regime): The Wall and its aftermath', in T.M. Wilson and H. Donnan (eds) *Border Identities: Nation and State at International Frontiers*, Cambridge: Cambridge University Press, 162–90.

Borsuk, R. (1991) 'Indonesian notes: Tourism year is seeking to set the maps straight', *Asian Wall Street Journal*, 28 June: 1, 8.

Bowden, J.J. (1959) 'The Texas–New Mexico boundary dispute along the Rio Grande', *Southwestern Historical Quarterly*, 63: 221–37.

Bowman, K.S. (1994) 'The border as locator and innovator of vice', *Journal of Borderlands Studies*, 9(1): 51–67.

Boyd, S.W. (1997) 'The impact of human disaster on tourism destination regions', paper presented at GAU International Tourism Conference on Challenged Tourism, Girne, Northern Cyprus, 3–7 December.

Boyd, S.W. (1999) 'North–south divide: The role of the border in tourism to Northern Ireland', *Visions in Leisure and Business*, 17(4): 50–71.

Bradbury, S.L. and Turbeville III, D.E. (1997) 'Communities in transition: The experience of towns on the Washington State/British Columbia border since the implementation of Free Trade', *Small Town*, 28(2): 10–15.

Bramwell, B. and Sharman, A. (1999) 'Collaboration in local tourism policymaking', *Annals of Tourism Research*, 26: 392–415.

Brant, M. and Gallagher, T.E. (1993) 'Tourism and the Old Order Amish', *Pennsylvania Folklife*, 43(2): 71–5.

Brewton, C. and Withiam, G. (1998) 'United States tourism policy: Alive but not well', *Cornell Hotel and Restaurant Administration Quarterly*, 39(1): 50–9.

Briner, H.J. (1991) 'Region Basiliensis: une région, trois pays, un avenir européen', *Bulletin, Association de Géographes Francais*, 5: 377–82.

Brohman, J. (1996) 'New directions in tourism for third world development', *Annals of Tourism Research*, 23: 48–70.

Broinowski, A. (ed.) (1982) *Understanding ASEAN*, New York: St Martin's Press.

Brooke, J. (1996) 'New checkpoint means freedom for prairie town', *The Globe and Mail* (Toronto), 22 January.

Brown, D. (1994) *The State and Ethnic Politics in Southeast Asia*, London: Routledge.

Brown, T.C. (1997) 'The fourth member of NAFTA: The U.S.–Mexico border', *Annals of the American Academy of Political and Social Science*, 550: 105–21.

BTC (1996 – copyright) *Baltic Sea Tourism Commission*, accessed 23 April 1998. http://www.balticsea.com

Buchholz, H. (1994) 'The inner-German border: Consequences of its establishment and abolition', in C. Grundy-Warr (ed.), *World Boundaries, Vol. 3, Eurasia*, London: Routledge, 55–62.

Buckley, P.J. and Witt, S.F. (1990) 'Tourism in the centrally-planned economies of Europe', *Annals of Tourism Research*, 17: 7–18.

Budd, J. (1990) 'Mexico's border towns to be focus of new tourism imperative', *Travel Weekly*, 49(49): 15.

Bumbaru, D. (1992) 'Dubrovnik: Heritage and culture as targets', *Impact*, 4(5): 1–2.

Burns, P. and Cooper, C. (1997) 'Yemen: Tourism and a tribal-Marxist dichotomy', *Tourism Management*, 18: 555–63.

Burton, R.C.J. (1994) 'Geographical patterns of tourism in Europe', *Progress in Tourism, Recreation and Hospitality Management*, 5: 3–25.

Buszynski, L. (1998) 'Thailand and Myanmar: The perils of "constructive engagement"', *The Pacific Review*, 11(2): 290–305.

Butler, R.W. (1993) 'Tourism development in small islands: Past influences and future directions', in D.G. Lockhart, D. Drakakis-Smith, and J. Schembri (eds) *The Development Process in Small Island States*, London: Routledge, 71–91.

Butler, R.W. (1996) 'The development of tourism in frontier regions: Issues and approaches', in Y. Gradus and H. Lithwick (eds) *Frontiers in Regional Development*, Lanham, MD: Rowman & Littlefield, 213–29.

Butler, R.W. and Mao, B. (1995) 'Tourism between quasi-states: International, domestic or what?', in R.W. Butler and D. Pearce (eds) *Change in Tourism: People, Places, Processes*, London: Routledge, 92–113.

Butler, R.W. and Mao, B. (1996) 'Conceptual and theoretical implications of tourism between partitioned states', *Asia Pacific Journal of Tourism Research*, 1(1): 25–34.

Bygvrå, S. (1990) 'Border shopping between Denmark and West Germany', *Contemporary Drug Problems*, 17(4): 595–611.

Cahill, R. (1987) *Border Towns of the Southwest: Shopping, Dining, Fun and Adventure from Tijuana to Juarez*, Boulder, CO: Pruett Publishing Co.

Campos, J.J., Bertenthal, B.I., and Kermoian, R. (1992) 'Early experience and emotional development: The emergence of wariness and heights', *Psychological Science*, 3: 61–4.

Canadian Chamber of Commerce (1992) *The Cross Border Shopping Issue*, Ottawa: Canadian Chamber of Commerce.

Canadian Press (1997a) 'Canadians to avoid border checks', *The Globe and Mail* (Toronto), 18 October.

Canadian Press (1997b) 'U.S. visa law is "long way off", official says', *Kitchener-Waterloo Record*, 14 October: 2.

Cao, X. and Ge, A. (1993) 'Shenzen: A new frontier' *China Tourism*, 156: 7–15.

Cárdenas, E.J. (1992) 'The Treaty of Asunción: A Southern Cone Common Market (Mercosur) begins to take shape', *World Competition*, 15(4): 65–77.

Carmichael, B.A., Peppard Jr., D.M. and Boudreau, F.A. (1996) 'Megaresort on my doorstep: Local resident attitudes toward Foxwoods Casino and casino gambling on nearby Indian reservation land', *Journal of Travel Research*, 34(3): 9–16.

Carpenter, W.C. (1925) 'The Red River boundary dispute (Oklahoma–Texas)', *American Journal of International Law*, 19: 517–29.

Carter, F.W. (1991) 'Bulgaria', in D.R. Hall (ed.) *Tourism and Economic Development in Eastern Europe and the Soviet Union*, London: Belhaven Press, 155–72.

Cartwright, D. (1988) 'Linguistic territorialization: Is Canada approaching the Belgian model?', *Journal of Cultural Geography*, 8(2): 115–34.

Cartwright, D. (1996) 'The expansion of French language rights in Ontario, 1968–1993: The uses of territoriality in a policy of gradualism', *Canadian Geographer*, 40: 238–57.

Castro, A. (1982) 'ASEAN economic co-operation', in A. Broinowski (ed.) *Understanding ASEAN*, New York: St Martin's Press, 70–91.

Catudal, H.M. (1979) *The Exclave Problem of Western Europe*, Tuscaloosa: University of Alabama Press.

Caviedes, C.N. (1994) 'Argentine–Chilean cooperation and disagreement along the southern Patagonian border', in W.A. Gallusser (ed.) *Political Boundaries and Coexistence*, Bern: Peter Lang, 135–43.

Chamberlain, L. (1991) *Small Business Ontario Report No. 44: Cross Border Shopping*, Toronto: Ministry of Industry, Trade and Technology.

Chardon, J.P. (1995) 'Saint-Martin ou l'implacable logique touristique', *Cahiers d'Outre-Mer*, 48(189): 21–33.

Chase, H. (1996) 'A road sign of good times: Sint Maarten/Saint Martin', *American Visions*, 11(6): 42.

Chatterjee, A. (1991) 'Cross-border shopping: Searching for a solution', *Canadian Business Review*, 18: 26–31.

Chenming, R. (1992) 'Cross-border one-day trips along the Chinese boundary in Yunnan', *China Tourism*, 145: 41–9.

China Radio International (1997) 'Republic of Korea reportedly to simplify visa procedures for PRC tourists', China Radio International broadcast, Beijing, 5 June.

Chosunilbo (1999) 'First foreigners go on Mt. Kumbang tour', *Chosunilbo*, 22 October.

Christaller, W. (1955) 'Beiträge zu einer Geographie des Fremdenverkehrs', *Erdkunde*, 9(1): 10–19.

Christaller, W. (1963) 'Some considerations of tourism in Europe: The peripheral regions-underdeveloped countries-recreation areas', *Papers of the Regional Science Association*, 12: 95–105.

Christian Science Monitor (1989) 'Israelis return last of captured land to Egypt', *Christian Science Monitor*, 16 March: 3.

Church, A. and Reid, P. (1995) 'Transfrontier co-operation, spatial development strategies and the emergence of a new scale of regulations: The Anglo-French border', *Regional Studies*, 29(3): 297–306.

Church, A. and Reid, P. (1996) 'Urban power, international networks and competition: The example of cross-border cooperation', *Urban Studies*, 33(8): 1297–318.

Ciaccio, C. (1979) 'L'organisation d'un espace périphérique: l'exemple de Pantellerie, limite méridional du tourisme en Italie', in G. Gruber, H. Lamping, W. Lutz, J. Matznetter, and K. Vorlaufer (eds) *Tourism and Borders: Proceedings of the Meeting of the IGU Working Group – Geography of Tourism and*

Recreation, Frankfurt: Institut für Wirtschafts- und Sozialgeographie der Johann Wolfgang Goethe Universität, 253–65.

Clark, T. (1994) 'National boundaries, border zones, and marketing strategy: A conceptual framework and theoretical model of secondary boundary effects', *Journal of Marketing*, 58: 67–80.

Clement, N., Ganster, P., and Sweedler, A. (1999) 'Development, environment, and security in asymmetrical border regions: European and North American perspectives', in H. Eskelinen, I. Liikanen, and J. Oksa (eds) *Curtains of Iron and Gold: Reconstructing Borders and Scales of Interaction*, Aldershot: Ashgate, 243–81.

Cohen, A. (1993) 'San Marino: The vest pocket republic has big ideas about its place in the world', *Destinations*, 8(2): 52–7.

Cole, J. and Cole, F. (1993) *The Geography of the European Community*, London: Routledge.

Comeaux, M.L. (1982) 'Attempts to establish and change a western boundary', *Annals of the Association of American Geographers*, 72: 254–71.

Conzen, M.P. (1990) 'Introduction', in M.P. Conzen (ed.), *The Making of the American Landscape*, Boston: Unwin Hyman, 1–8.

Cosaert, P. (1994) 'Frontiere et commerce de détail: la localisation des commerces de détail aux points de passage de la frontiére franco-belge au niveau de l'arrondissement de Lille', *Hommes et Terres du Nord*, 2(3): 134–41.

Cowell, A. (1997) 'Austrian border controls are slow to fall', *New York Times*, 20 July: 3.

Cowley, J. and Lemon, A. (1986) 'Bophuthatswana: Dependent development in a black "homeland"', *Geography*, 71(3): 252–5.

Crush, J. and Wellings, P. (1983) 'The southern African pleasure periphery, 1966–1983', *Journal of Modern African Studies*, 21(4): 673–98.

Crush, J. and Wellings, P. (1987) 'Forbidden fruit and the export of vice: Tourism in Lesotho and Swaziland', in S. Britton and W.C. Clarke (eds) *Ambiguous Alternative: Tourism in Small Developing Countries*, Suva, Fiji: University of the South Pacific, 91–112.

Cullen, K. (1998) 'Britain and Ireland unveil draft for an Ulster accord', *Boston Globe*, 13 January.

Cummings, J. (1992) *Thailand: A Travel Survival Kit*, Fifth Edition, Hawthorn, Australia: Lonely Planet.

Curtis, J.R. (1993) 'Central business districts of the two Loredos', *Geographical Review*, 83(1): 54–65.

Curtis, J.R. and Arreola, D.D. (1989) 'Through Gringo eyes: Tourist districts in the Mexican border cities as other-directed places', *North American Culture*, 5(2): 19–32.

Curtis, J.R. and Arreola, D.D. (1991) 'Zonas de tolerancia on the northern Mexican border', *Geographical Review*, 81(3): 333–46.

D'Amore, L.J. (1983) 'Guidelines to planning in harmony with the host community', in P.E. Murphy (ed.) *Tourism in Canada: Selected Issues and Options*, Victoria, BC: University of Victoria, Department of Geography, 135–59.

de Albuquerque, K. and McElroy, J.L. (1995) 'Tourism development in small islands: St Maarten/St Martin and Bermuda', in D. Barker and D.F.M. McGregor (eds) *Environment and Development in the Caribbean: Geographical Perspectives*, Kingston: University of the West Indies Press, 70–89.

de Kadt, E. (1979) 'Social planning for tourism in the developing countries', *Annals of Tourism Research*, 6: 36–48.

de Vorsey Jr., L. (1982) *The Georgia–South Carolina Boundary: A Problem in Historical Geography*, Athens, GA: University of Georgia Press.

Delaware Tourism Office (1987) *Delaware: Small Wonder*, Dover: Delaware Tourism Office.

Denisiuk, Z., Stoyko, S., and Terray, J. (1997) 'Experience in cross-border cooperation for national park and protected areas in central Europe', in J.G. Nelson and R. Serafin (eds) *National Parks and Protected Areas: Keystones to Conservation and Sustainable Development*, Berlin: Springer, 145–50.

Denver Post (1995) 'Wan reception for U.S. immigration proposal', *Denver Post*, 30 August: A9.

DeQuine, J. (1989) 'Spring breakers head for the border', *USA Today*, 13 March: 1, 14.

Detroit News (1997) 'Fun at your fingertips! Casino Windsor', *Detroit News*, 31 August: 15A.

Dewailly, J.M. (1979) 'Fascination et pesanteur d'une frontière pour le tourisme et la récréation l'example Franco-Belge', in Gruber, H. Lamping, W. Lutz, J. Matznetter, and K. Vorlaufer (eds) *Tourism and Borders: Proceedings of Meeting of the IGU Working Group - Geography of Tourism and Recreation*, Frankfurt: Institut für Wirtschafts- und Sozialgeographie der Johann Wolfgang Goethe Universität, 309–17.

Di Matteo, L. (1993) 'Determinants of cross-border trips and spending by Canadians in the United States: 1979–1991', *Canadian Business Economics*, 1(3): 51–61.

Di Matteo, L. (1999) 'Cross-border trips by Canadians and Americans and the differential impact of the border', *Visions in Leisure and Business*, 17(4): 72–92.

Di Matteo, L. and Di Matteo, R. (1993) 'The determinants of expenditures by Canadian visitors to the United States', *Journal of Travel Research*, 31(4): 34–42.

Di Matteo, L. and Di Matteo, R. (1996) 'An analysis of Canadian cross-border travel', *Annals of Tourism Research*, 23: 103–22.

Diehl, P.N. (1983) 'The effects of the peso devaluation on Texas border cities', *Texas Business Review*, 57: 120–5.

Dieke, P.U.C. (1998) 'Regional tourism in Africa: scope and critical issues', in E. Laws, B. Faulkner, and G. Moscardo (eds) *Embracing and Managing Change in Tourism: International Case Studies*, London: Routledge, 29–48.

Diggines, C.E. (1985) 'The problems of small states', *The Round Table*, 295: 191–205.

Dilley, R.S. and Hartviksen, K.R. (1993) 'Duluth and Thunder Bay tourism after the Free Trade Agreement', *Operational Geographer*, 11(3): 15–18.

Dilley, R.S., Hartviksen, K.R., and Nord, D.C. (1991) 'Duluth and Thunder Bay: A study of mutual tourist attractions', *Operational Geographer*, 9(4): 9–13.

Dillman, C.D. (1970a) 'Recent developments in Mexico's National Border Program', *Professional Geographer*, 22(5): 243–7.

Dillman, C.D. (1970b) 'Urban growth along Mexico's northern border and the Mexican National Border Program', *Journal of Developing Areas*, 4: 487–507.

Dillman, C.D. (1976) 'Maquiladoras in Mexico's northern border industrialization program', *Tijdschrift voor Economische en Sociale Geografie*, 67: 138–50.

Direction du Tourisme (1994) *Monte Carlo: Places of Interest*, Monte Carlo: Direction du Tourisme et des Congrès.

do Amaral, I. (1994) 'New reflections on the theme of international boundaries', in C.H. Schofield (ed.) *World Boundaries, Vol. 1, Global Boundaries*, London: Routledge, 16–22.

Dobson, W.J. and Fravel, M.T. (1997) 'Red herring hegemon: China in the South China Sea', *Current History*, 96: 258–63.

Donaldson, G. (1979) *Niagara! The Eternal Circus*, Toronto: Doubleday Canada.

Donnan, H. and Wilson, T.M. (1999) *Borders: Frontiers of Identity, Nation and State*, Oxford: Berg.

Dorsey, J. (1990) 'Tourism council urges governments to drop barriers to travel', *Travel Weekly*, 49(17): 6.

Douglas, N. (1998) 'The politics of accommodation, social change and conflict resolution in Northern Ireland', *Political Geography*, 17: 209–30.

Drysdale, A. (1991) 'The Gulf of Aqaba coastline: An evolving border landscape', in D. Rumley and J.V. Minghi (eds) *The Geography of Border Landscapes*, London: Routledge, 203–16.

Dupuy, P.M. (1982) 'Legal aspects of transfrontier regional co-operation', *West European Politics*, 5: 50–63.

Dykstra, T.L. and Ironside, R.G. (1972) 'The effects of the division of the city of Lloydminster by the Alberta–Saskatchewan inter-provincial boundary', *Cahiers de Géographie de Québec*, 16: 263–83.

Eadington, W.R. (ed.) (1990) *Indian Gaming and the Law*, Reno: Institute for the Study of Gambling and Commercial Gaming, University of Nevada.

Eadington, W.R. (1996) 'The legalization of casinos: Policy objectives, regulatory alternatives, and cost/benefit considerations', *Journal of Travel Research*, 34(3): 3–8.

Eberstadt, N. (1995) *Korea Approaches Reunification*, Armonk, NY: M.E. Sharpe.

Echikson, W. (1997) 'A new map for an old continent: EU picks 11 nations to be partners', *Christian Science Monitor*, 15 December: 7.

Economic and Business Review Indonesia (1993) 'Cross-border shopping craze', *Economic and Business Review Indonesia*, 57: 34.

Economist (1989) 'Bouncing the beach ball', *The Economist*, 25 February: 38.

Economist (1991) 'On the borderline', *The Economist*, 23 November: 58.

Economist (1992) 'The future, and by and large it works', *The Economist*, 27 June: 61.

Economist (1993a) 'Coming of age in Andorra', *The Economist*, 13 March: 60.

Economist (1993b) 'Propinquity pays', *The Economist*, 20 February: 67.

Economist (1995) 'Red Sea Riviera?', *The Economist*, 16 December: 58.

Economist (1997) 'Berliners see red', *The Economist*, 8 March: 56.

Economist Intelligence Unit (1995) *International Tourism Reports: France*, London: Economist Intelligence Unit.

Edwards, J.N., Fuller, T.D., and Vorakitphokatorn, S. (1994) 'Why people feel crowded: An examination of objective and subjective crowding', *Population and Environment*, 16(2): 149–60.

Egerö, B. (1991) *South Africa's Bantustans: From Dumping Grounds to Battlefronts*, Uppsala, Sweden: Nordiska Afrikainstitutet.

El-Agraa, A.M. (1990) 'The theory of economic integration', in A.M. El-Agraa (ed.) *The Economics of the European Community*, New York: St Martin's Press, 79–96.

Elkins, T.H. and Hofmeister, B. (1988) *Berlin: The Spatial Structure of a Divided City*, London: Methuen.

Ellger, C. (1992) 'Berlin: Legacies of division and problems of unification', *Geographical Journal*, 158(1): 40–6.

Elliott, M. (1997) 'Hey, can you spare a "Euro"?', *Newsweek*, 17 February: 48–9.

Ellman, E. and Robbins, D. (1998) 'Merging sustainable development with wastewater infrastructure improvement on the U.S.–Mexico border', *Journal of Environmental Health*, 60(7): 8–13.

English Heritage (1996) *Hadrian's Wall World Heritage Site Management Plan*, London: English Heritage.

Eriksson, G.A. (1979) 'Tourism at the Finnish–Swedish–Norwegian borders', in G. Gruber, H. Lamping, W. Lutz, J. Matznetter, and K. Vorlaufer (eds) *Tourism and Borders: Proceedings of the Meeting of the IGU Working Group – Geography of Tourism and Recreation*, Frankfurt: Institut für Wirtschafts- und Sozialgeographie der Johann Wolfgang Goethe Universität, 151–62.

Essex, S.J. and Gibb, R.A. (1989) 'Tourism in the Anglo-French frontier zone', *Geography*, 74(3): 222–31.

European Central Bank (1999) *The Euro Banknotes and Coins*, Frankfurt: European Central Bank.

European Commission (1995) *The Role of the Union in the Field of Tourism*, Brussels: Commission of the European Communities.

Fagence, M. (1997) 'An uncertain future for tourism in microstates: The case of Nauru', *Tourism Management*, 18: 385–92.

Felsenstein, D. and Freeman, D. (1998) 'Simulating the impacts of gambling in a tourist location: Some evidence from Israel', *Journal of Travel Research*, 37(2): 145–55.

Fernandez, R.A. (1977) *The United States–Mexico Border: A Politico-Economic Profile*, Notre Dame: University of Notre Dame Press.

Fesenmaier, D.R. and Vogt, C.A. (1993) 'Evaluating the economic impact of travel information provided at Indiana welcome centers', *Journal of Travel Research*, 31(3): 33–9.

Fesenmaier, D.R., Vogt, C.A., and Stewart, W.P. (1993) 'Investigating the influence of welcome center information on travel behavior', *Journal of Travel Research*, 31(3): 47–52.

Fineberg, A. (1993) *Regional Cooperation in the Tourism Industry*, Jerusalem: Israel/Palestine Center for Research and Information.

Finn, P. (2000) 'Berlin seeking to keep memory of the Wall alive', *Arizona Republic*, 27 August: A24.

Fisher, M. (1990) 'Canadians give in to shopping drive', *The Globe and Mail* (Toronto), 26 October: A4.

Fitzgerald, J.D., Quinn, T.P., Whelan, B.J., and Williams, J.A. (1988) *An Analysis of Cross-Border Shopping*, Dublin: The Economic and Social Research Institute.

Fordney, C. (1996) 'Boundary wars', *National Parks*, 70(1): 24–9.

Foster-Carter, A. (1981) 'North Korea: Opening the door', *New Society*, 27 August: 224.

Foster-Carter, A. (1994) 'Korea: Sociopolitical realities of reuniting a divided nation', in T.H. Henriksen and K. Lho (eds) *One Korea? Challenges and Prospects for Reunification*, Stanford, CA: Hoover Institution Press, Stanford University, 31–47.

Foster-Carter, A. (1996) 'Monumental puzzle', *Far Eastern Economic Review*, 26 June: 36–9.

Fox, W.F. (1986) 'Tax structure and the location of economic activity along state borders', *National Tax Journal*, 39(4): 387–401.

Friedmann, J. (1966) *Regional Development Policy: A Case of Venezuela*, Cambridge, MA: MIT Press.

Friedmann, J. (1992) *Empowerment: The Politics of Alternative Development*, Oxford: Blackwell.

Friedman, J.S. (1988) 'Taba goes to Egypt: Settlement of border dispute with Israel boosts Egypt's leader', *Christian Science Monitor*, 30 September: 11.

Gabe, T., Kinsey, J., and Loveridge, S. (1996) 'Local economic impacts of tribal casinos: The Minnesota case', *Journal of Travel Research*, 34(4): 81–8.

Gaines, S.E. (1995) 'Bridges to a better environment: Building cross-border institutions for environmental improvement in the U.S.–Mexico border area', *Arizona Journal of International and Comparative Law*, 12(2): 429–71.

Gebhardt, H. (1987) 'Perzeption von Grenzen und Grenzuberschreitende Verflechtungen Beispiele aus dem Alpenraum', *Revue Geographique de l'Est*, 27(1): 39–51.

Geldenhuys, D. (1990) *Isolated States: A Comparative Analysis*, Cambridge: Cambridge University Press.

Gengxin, J. (1997) 'Heilongjiang achieves gratifying results in border trade', *Heilongjiang Ribao*, 15 May: 2.

Geographical Magazine (1992) 'Greenwich lean time', *Geographical Magazine*, 64(8): 5.

Gerlach, J. (1989) 'Spring break at Padre Island: A new kind of tourism', *Focus*, 39(1): 13–16, 29.

German, A.L. (1984) 'Point Roberts: A tiny borderline anomaly', *Canadian Geographic*, 104(5): 72–4.

Getz, D. (1987) 'Tourism planning and research: Traditions, models and futures', in *Proceedings of the Australian Travel Workshop*, Bunbury, Western Australia: Australian Travel Workshop, 407–48.

Getz, D. (1992) 'Tourism planning and destination life cycle', *Annals of Tourism Research*, 19: 752–70.

Getz, D. (1993) 'Planning for tourism business districts', *Annals of Tourism Research*, 20(3): 583–600.

Gibb, R. (1985) 'No-passport excursions to France: A case study of tension management', *Area*, 17(2): 89–96.

Gibbons, J.D. and Fish, M. (1987) 'Market sensitivity of U.S. and Mexican border travel', *Journal of Travel Research*, 26(1): 2–6.

Gibson, B. (1998) 'Letter to Planet Talk', *Planet Talk*, 33: 3.

Gitelson, R. and Perdue, R.R. (1987) 'Evaluating the role of state welcome centers in disseminating travel related information in North Carolina', *Journal of Travel Research*, 25(4): 15–19.

Glassner, M.I. (1996) *Political Geography*, Second Edition, New York: Wiley.

Gonin, P. (1994) 'Régions frontalières et développement endogène: de nouveaux territoires en construction au sein de l'Union Européenne', *Hommes et Terres du Nord*, 2(3): 61–70.

Gonzales Associates (1970) *Proposed General Plan, City of Nogales, Arizona, Nogales Downtown Area*, Phoenix: Gonzales Associates.

Goodall, B. (1987) *Dictionary of Human Geography*, New York: Facts on File Publications.

Goodman, L.R. (1992) 'A working paper on crossborder shopping: The Canadian impact on North Dakota', in H.J. Selwood and J.C. Lehr (eds) *Reflections from the Prairies: Geographical Essays*, Winnipeg: University of Winnipeg, Department of Geography, 80–9.

Gormsen, E. (1979) 'The impact of tourism on the development of Mexican cities along the U.S. border: The example of Tijuana', in G. Gruber, H. Lamping, W. Lutz, J. Matznetter, and K. Vorlaufer (eds) *Tourism and Borders: Proceedings of the Meeting of the IGU Working Group – Geography of Tourism and Recreation*, Frankfurt: Institut für Wirtschafts- und Sozialgeographie der Johann Wolfgang Goethe Universität, 345.

Gormsen, E. (1995) 'International tourism in China: Its organization and socio-economic impact', in A.A. Lew and L. Yu (eds) *Tourism in China: Geographic, Political, and Economic Perspectives*, Boulder, CO: Westview Press, 63–88.

Gosar, A. (1999) 'Recovering tourism in the Balkans', paper presented at the Annual Meetings of the Association of American Geographers, Honolulu, 23 March.

Gosar, A. and Klemenčič, V. (1994) 'Current problems of border regions along the Slovene–Croatian border', in W.A. Gallusser (ed.) *Political Boundaries and Coexistence*, Bern: Peter Lang, 30–42.

Gottman, J. (1973) *The Significance of Territory*, Charlottesville: University Press of Virginia.

Government of Israel (1997) *Programs for Regional Cooperation*, Tel Aviv: Ministry of Foreign Affairs and Ministry of Finance.

Government of New Brunswick (1992) *A Discussion Paper on Cross Border Shopping*, Fredericton: Department of Economic Development and Tourism.

Government of Ontario (1991) *Report on Cross-Border Shopping*, Toronto: Standing Committee on Finance and Economic Affairs.

Gradus, Y. (1994) 'The Israel–Jordan Rift Valley: A border of cooperation and productive coexistence', in W.A. Gallusser (ed.), *Political Boundaries and Coexistence*, Bern: Peter Lang, 315–21.

Graf, W.D. (1992) 'Sustainable ideologies and interests: Beyond Brundtland', *Third World Quarterly*, 13: 553–9.

Graizbord, C. (1986) 'Trans-boundary land-use planning: A Mexican perspective', in L.A. Herzog (ed.) *Planning the International Border Metropolis: Trans-Boundary Policy Options in the San Diego–Tijuana Region*, San Diego: Center for U.S.–Mexican Studies, University of California, 13–20.

Greene, B.M. (1996) 'The reservation gambling fury: Modern Indian uprising or unfair restraint on tribal sovereignty?', *BYU Journal of Public Law*, 10(1): 93–116.

Greenhouse, L. (1998) 'High court awards New Jersey sovereignty over most of Ellis Island', *New York Times*, 27 May: 1, 21.

Greenwich Waterfront Development Partnership (1995) *Greenwich 2000: Tourism Development*, Greenwich: Waterfront Development Partnership.

Greenwich Waterfront Development Partnership (1996) *Once in a Thousand Years: Maximising Opportunities in Greenwich*, Greenwich: Waterfront Development Partnership.

Griffin, E.C. and Ford, L.R. (1980) 'Model of Latin American city structure', *Geographical Review*, 70: 397–422.

Griffith, P. (1997) 'New law threatens gridlock on border', *The Blade* (Toledo), 28 September: 1, 4.

Grinspun, R. (1993) 'The economics of free trade in Canada', in R. Grinspun and M. Cameron (eds) *The Political Economy of North American Free Trade*, New York: St Martin's Press, 105–24.

Griswold, E.N. (1939) 'Hunting boundaries with car and camera in the Northeastern United States', *Geographical Review*, 29: 353–82.

Groth, P. and Bressi, T.W. (eds) (1997) *Understanding Ordinary Landscapes*, New Haven: Yale University Press.

Gruber, G.R. (1979) 'The influence of national borders on tourism in Africa (the Zambian example)', in G.R. Gruber, H. Lamping, W. Lutz, J. Matznetter, and K. Vorlaufer (eds) *Tourism and Borders: Proceedings of the Meeting of the IGU Working Group – Geography of Tourism and Recreation*, Frankfurt: Institut für Wirtschafts- und Sozialgeographie der Johann Wolfgang Goethe Universität, 181–94.

Gruber, G., Lamping, H., Lutz, W., Matznetter, J., and Vorlaufer, K. (eds) (1979) *Tourism and Borders: Proceedings of the Meeting of the IGU Working Group – Geography of Tourism and Recreation*, Frankfurt: Institut für Wirtschafts- und Sozialgeographie der Johann Wolfgang Goethe Universität.

Grundy-Warr, C. (1994) 'Peacekeeping lessons from divided Cyprus', in C. Grundy-Warr (ed.) *World Boundaries, Vol. 3, Eurasia*, London: Routledge, 71–88.

Grundy-Warr, C. and Schofield, R.N. (1990) 'Man-made lines that divide the world', *Geographical Magazine*, 62(6): 10–15.

Gunn, C.A. (1994) *Tourism Planning: Basics, Concepts, Cases*, Third Edition, Washington, DC: Taylor and Francis.

Hagiwara, Y. (1973) 'Formation and development of the Association of Southeast Asian Nations', *The Developing Economies*, 11(4): 443–65.

Hajdú, Z. (1994) 'Cities in the frontier regions of the Hungarian state: Changing borders, changing political systems, new evaluated contacts', in W.A. Gallusser (ed.) *Political Boundaries and Coexistence*, Bern: Peter Lang, 209–18.

Hall, C.M. (1994a) 'The closer economic relationship between Australia and New Zealand: Implications for travel and tourism', *Journal of Travel and Tourism Marketing*, 3: 123–31.

Hall, C.M. (1994b) *Tourism and Politics: Policy, Power and Place*, Chichester: Wiley.

Hall, C.M. and Page, S.J. (eds) (2000) *Tourism in South and Southeast Asia: Critical Perspectives*, Oxford: Butterworth-Heinemann.

Hall, D.R. (1984) 'Foreign tourism under socialism: The Albanian "Stalinist" model', *Annals of Tourism Research*, 11: 539–55.

Hall, D.R. (1986a) 'North Korea opens to tourism: A last resort', *Inside Asia*, 9(4): 21–3.

Hall, D.R. (1986b) 'North of the divide', *Geographical Magazine*, 58(11): 590–2.

Hall, D.R. (1990a) 'Eastern Europe opens its doors', *Geographical Magazine*, 62(4): 10–15.

Hall, D.R. (1990b) 'Stalinism and tourism: A study of Albania and North Korea', *Annals of Tourism Research*, 17: 36–54.

Hall, D.R. (1990c) 'The "communist world" in the 1990s', *Town & Country Planning*, 59(1): 28–30.

Hall, D.R. (1991a) 'Evolutionary pattern of tourism development in Eastern Europe and the Soviet Union', in D.R. Hall (ed.) *Tourism and Economic Development in Eastern Europe and the Soviet Union*, London: Belhaven Press, 79–115.

Hall, D.R. (1991b) 'Introduction', in D.R. Hall (ed.) *Tourism and Economic Development in Eastern Europe and the Soviet Union*, London: Belhaven Press, 3–28.

Hall, D.R. (1992) 'Albania's changing tourism environment', *Journal of Cultural Geography*, 12(2): 35–44.

Hall, D.R. (1995) 'Tourism change in Central and Eastern Europe', in A. Montanari and A.M. Williams (eds) *European Tourism: Regions, Spaces and Restructuring*, Chichester: Wiley, 221–44.

Hamilton, J. and Broustas, M.V. (1996) 'Peace dividend: Tourists reclaim the Balkans', *Traveler*, November: 37.

Hansen, N. (1977) 'The economic development of border regions', *Growth and Change*, 8(4): 2–8.

Hansen, N. (1981) *The Border Economy: Regional Development in the Southwest*, Austin: University of Texas Press.

Hansen, N. (1983) 'International cooperation in border regions: An overview and research agenda', *International Regional Science Review*, 8(3): 255–70.

Hansen, N. (1986) 'Border region development and cooperation: Western Europe and the U.S.–Mexico borderlands in comparative perspective', in O.J. Martinez (ed.) *Across Boundaries: Transborder Interaction in Comparative Perspective*, El Paso: Center for Inter-American and Border Studies, University of Texas, 31–44.

Harris, C.D. (1991) 'Unification of Germany in 1990', *Geographical Review*, 81: 170–82.

Harris, M. (1997) 'Spirited, independent Slovenia', *New York Times*, 11 May: 12.

Harrison, D. (1992) 'Tradition, modernity and tourism in Swaziland', in D. Harrison (ed.) *Tourism and the Less Developed Countries*, Chichester: Wiley, 148–62.

Hartshorne, R. (1936) 'Suggestions on the terminology of political boundaries', *Annals of the Association of American Geographers*, 26: 56–7.

Hasson, S. and Razin, E. (1990) 'What is hidden behind a municipal boundary conflict?', *Political Geography Quarterly*, 9(3): 267–83.

Haywood, K.M. (1988) 'Responsible and responsive tourism planning in the community', *Tourism Management*, 9: 105–18.

Hazan, R. (1988) 'Peaceful conflict resolution in the Middle East: The Taba negotiations', *Journal of the Middle East Studies Society*, 2(1): 39–65.

Helle, R.K. (1979) 'Observations on tourism between Finland and the Soviet Union', in G. Gruber, H. Lamping, W. Lutz, J. Matznetter, and K. Vorlaufer (eds) *Tourism and Borders: Proceedings of the Meeting of the IGU Working Group – Geography of Tourism and Recreation*, Frankfurt: Institut für Wirtschafts- und Sozialgeographie der Johann Wolfgang Goethe Universität, 163–7.

Henderson, L. (1993) 'Forging a link: Two approaches to integrating trade and environment', *Alternatives*, 20(1): 30–6.

Henriksen, T.H. and Lho, K. (1994) 'Introduction', in T.H. Henriksen and K. Lho (eds) *One Korea? Challenges and Prospects for Reunification*, Stanford, CA: Hoover Institution Press, Stanford University, 1–11.

Herrick, T. (1997) 'Blurring the line', *Houston Chronicle*, 10 August: D1.

Herzog, L.A. (1982) 'Cross cultural barriers to planned urban development in the U.S.–Mexico border zone: A case study of the San Diego/Tijuana metropolitan region', in E.M. Berrueto (ed.) *Proceedings of the First Conference on Regional Impacts of United States–Mexico Economic Relations*, Guanajuato, Mexico: Conference on Regional Impacts of United States–Mexico Economic Relations, 836–67.

Herzog, L.A. (1985) 'Tijuana', *Cities*, 2: 297–306.

Herzog, L.A. (1986a) 'Overview', in L.A. Herzog (ed.) *Planning the International Border Metropolis: Trans-Boundary Policy Options in the San Diego–Tijuana Region*, San Diego: Center for U.S.–Mexican Studies, University of California, 67–71.

Herzog, L.A. (1986b) 'San Diego–Tijuana: The emergence of a trans-boundary metropolitan ecosystem', in L.A. Herzog (ed.) *Planning the International Border Metropolis: Trans-Boundary Policy Options in the San Diego–Tijuana Region*, San Diego: Center for U.S.–Mexican Studies, University of California, 1–10.

Herzog, L.A. (1990) *Where North Meets South: Cities, Space, and Politics on the U.S.–Mexico Border*, Austin: Center for Mexican American Studies, University of Texas.

Herzog, L.A. (1991a) 'Cross-national urban structure in the era of global cities: The U.S.–Mexico transfrontier metropolis', *Urban Studies*, 28(4): 519–33.

Herzog, L.A. (1991b) 'The transfrontier organization of space along the U.S.–Mexico border', *Geoforum*, 22(3): 255–69.

Herzog, L.A. (1991c) 'USA–Mexico border cities: A clash of two cultures', *Habitat International*, 15(1/2): 261–73.

Herzog, L.A. (1992) 'The U.S.–Mexico transfrontier metropolis', *Business Mexico*, 2: 14–17.

Heung, V.C.S. (1997) 'Hong Kong: Political impact on tourism', in F.M. Go and C.L. Jenkins (eds) *Tourism and Economic Development in Asia and Australasia*, London: Cassell, 123–37.

Hidalgo, L. (1993) 'British shops suffer as "booze cruise" bargain hunters flock to France', *The Times*, 22 November: 5.

Ho, K.C. and So, A. (1997) 'Semi-periphery and borderland integration: Singapore and Hong Kong experiences', *Political Geography*, 16(3): 241–59.

Hobson, J.S.P. (1995) 'Hong Kong: The transition to 1997', *Tourism Management*, 16: 15–20.

Hocking, B. and McGuire, S. (eds) (1999) *Trade Politics: International, Domestic, and Regional Perspectives*, London: Routledge.

Hodgkinson, T. (1992) 'Ottawa to launch travel campaign', *London Free Press*, 5 September.

Høivik, T. and Heiberg, T. (1980) 'Centre-periphery tourism and self-reliance', *International Social Science Journal*, 32(1): 69–98.

Holden, R.J. (1984) 'Maquiladoras' employment and retail sales effects on four Texas border communities, 1978–1983', *Southwest Journal of Business and Economics*, 2(1): 16–26.

Holdich, T.H. (1916) *Political Frontiers and Boundary Making*, New York: Macmillan.

Hoon, S.J. (1990) 'They shall not pass: Efforts to allow cross-border visits collapse', *Far Eastern Economic Review*, 23 August: 11.

Hoseason, J. (1998) 'The Euro affair: How a single currency for Europe will affect tourism', *Tourism*, 96: 10–11.

House, J.W. (1980) 'The frontier zone: A conceptual problem for policy makers', *International Political Science Review*, 1(4): 456–77.

House, J.W. (1981) 'Frontier studies: An applied approach', in A.D. Burnett and P.J. Taylor (eds) *Political Studies from Spatial Perspectives*, New York: Wiley, 291–312.

Hovinen, G.R. (1995) 'Heritage issues in urban tourism: An assessment of new trends in Lancaster County', *Tourism Management*, 16: 381–8.

Howard, D. and Gitelson, R. (1989) 'An analysis of the differences between state welcome center users and nonusers: A profile of Oregon vacationers', *Journal of Travel Research*, 28(4): 38–40.

Hudman, L.E. (1978) 'Tourist impacts: The need for planning', *Annals of Tourism Research*, 5: 112–25.

Hudson, B.M. (1979) 'Comparison of current planning theories: Counterparts and contradictions', *Journal of the American Planning Association*, 45: 387–98.

Hufbauer, G.C. and Schott, J.J. (1992) *North American Free Trade: Issues and Recommendations,* Washington, DC: Institute for International Economics.

Huntoon, L. (1998) 'Immigration to Spain: Implications for a unified European Union immigration policy', *International Migration Review*, 32(2): 423–50.

Husbands, W. (1981) 'Centres, peripheries, tourism and socio-spatial development', *Ontario Geography*, 17: 37–59.

Hussey, A. (1991) 'Regional development and cooperation through ASEAN', *Geographical Review*, 81: 87–98.

Ingram, H., Milich, L., and Varady, R.G. (1994) 'Managing transboundary resources: Lessons from Ambos Nogales', *Environment*, 36(4): 6–9, 28–38.

Inskeep, E. (1991) *Tourism Planning: An Integrated and Sustainable Development Approach*, New York: Van Nostrand Reinhold.

International Peace Garden (n.d.) *Like No Place on Earth!,* Boissevain, MB: International Peace Garden.

Ioannides, D. (1992) 'Tourism development agents: The Cypriot resort cycle', *Annals of Tourism Research*, 19: 711–31.

Ioannides, D. and Apostolopoulos, Y. (1999) 'Political instability, war, and tourism in Cyprus: Effects, management, and prospects for recovery', *Journal of Travel Research*, 38(1): 51–6.

Iseminger, G. (1991) 'Stone border', *North Dakota Horizons*, 21(1): 14–21.

Jackson, J.B. (1984) *Discovering the Vernacular Landscape,* New Haven: Yale University Press.

Jackson, R.H. and Hudman, L.E. (1987) 'Border towns, gambling and the Mormon culture region', *Journal of Cultural Geography*, 8(1): 35–48.

Jamal, T.B. and Getz, D. (1995) 'Collaboration theory and community tourism planning', *Annals of Tourism Research*, 22: 186–204.

Jansen-Verbeke, M.C. (1990) 'From leisure shopping to shopping tourism', in *Proceedings of the International Sociological Association Annual Conference*, Madrid: International Sociological Association, 1–17.

Jenkins, L. (1988) 'San Marino', *The Washington Post*, 19 June: 13–14.

Jenner, P. and Smith, C. (1993) 'Europe's microstates: Andorra, Monaco, Liechtenstein and San Marino', *EIU International Tourism Reports*, 1: 69–89.

Jessop, B. (1995) 'Regional economic blocs, cross-border cooperation, and local economic strategies in postcolonialism', *American Behavioral Scientist*, 38(5): 674–715.

Johnston, M., Mauro, R., and Dilley, R.S. (1991) 'Trans-border tourism in a regional context', paper presented at the annual meeting of the Canadian Association of Geographers, Kingston, Ontario.

Johnston, R.J. (1982) *Geography and the State: An Essay in Political Geography,* New York: St Martin's Press.

Johnstone, N. (1995) 'International trade, transfrontier pollution, and environmental cooperation: A case study of the Mexican–American border region', *Natural Resources Journal*, 35(1): 33–62.

Jones, K. (1997) 'Slipping in and out of Mexico', *New York Times*, 26 January: 15.

Jones, S.B. (1943) 'The description of international boundaries', *Annals of the Association of American Geographers*, 49: 241–55.

Jones, S.B. (1945) *Boundary Making*, Washington, DC: Carnegie Endowment.

Joyce, J. (1990) 'Court rules for border bargains', *Guardian*, 13 June.

Jud, G.D. (1975) 'Tourism and crime in Mexico', *Social Science Quarterly*, 56(2): 324–30.

Kaluski, S. (1994) 'The "new" eastern Polish border in the face of environmental and socio-political problems', in W.A. Gallusser (ed.) *Political Boundaries and Coexistence*, Bern: Peter Lang, 57–64.

Kammas, M. (1991) 'Tourism development in Cyprus', *The Cyprus Review*, 3(2): 7–26.

Kamrava, M. (1993) *Politics and Society in the Third World*, London: Routledge.

Kearney, E.P. (1992) 'Redrawing the political map of tourism: The European view', *Tourism Management*, 13: 34–6.

Keller, C.P. (1984) 'Centre-periphery tourism development and control', in J. Long and R. Hecock (eds) *Leisure, Tourism and Social Change*, Edinburgh: Centre for Leisure Research, 77–84.

Keller, C.P. (1987) 'Stages of peripheral tourism development – Canada's Northwest Territories', *Tourism Management*, 8: 20–32.

Kemp, A. and Ben-Eliezer, U. (2000) 'Dramatizing sovereignty: The construction of territorial dispute in the Israeli–Egyptian border at Taba', *Political Geography*, 19: 315–44.

Kemp, K. (1992) 'Cross-border shopping: Trends and measurement issues', *Canadian Economic Observer*, 5: 1–13.

Kendall, K.W. and Kreck, L.A. (1992) 'The effect of the across-the-border travel of Canadian tourists on the city of Spokane: A replication', *Journal of Travel Research*, 30(4): 53–8.

Kenney, J. (1991) 'Beyond park boundaries: The spoils of development and industry imperil nearby national parks', *National Parks*, 65(7/8): 20–5.

Keown, I. (1991) 'Border town: Stay in Holland's medieval Maastricht and see four countries in one unhurried day', *Travel & Leisure*, 21(4): 112–21.

Kersell, J.E. (1991) 'Political fragmentation of the Dutch Windward Islands', *Public Administration and Development*, 11: 79–88.

Kersell, J.E., Brookson, A., Duzanson, L.L., Groeneveldt, R.A., and Arts, X. (1993) 'Small-scale administration in St Martin: Two governments of one people', *Public Administration and Development*, 13: 49–64.

Kiefer, F.S. (1989) 'Berliners rejoice at open gate', *Christian Science Monitor*, 26 December: 3.

Kim, Y.K. and Crompton, J.L. (1990) 'Role of tourism in unifying the two Koreas', *Annals of Tourism Research*, 17: 353–66.

Kinzer, S. (1994) 'At Checkpoint Charlie, a museum remembers', *New York Times*, 18 December: 3.

Kirby, K.M. (1996) *Indifferent Boundaries: Spatial Concepts of Human Subjectivity*, New York: Guildford Press.

Kirk, W. (1963) 'Problems of geography', *Geography*, 48: 357–71.

Kissling, C. (1998) 'Beyond the Australasian single aviation market', *Australian Geographical Studies*, 36(2): 170–6.

Kitchener-Waterloo Pennysaver (1993) 'Shopping trip to Erie, Pennsylvania', *Kitchener-Waterloo Pennysaver*, 31 October.

Kiy, R. and Wirth, J.D. (1998) 'Introduction', in R. Kiy and J.D. Wirth (eds) *Environmental Management on North America's Borders*, College Station: Texas A&M University Press, 3–31.

Kjos, K. (1986) 'Trans-boundary land-use planning: A view from San Diego County', in L.A. Herzog (ed.) *Planning the International Border Metropolis: Trans-Boundary Policy Options in the San Diego–Tijuana Region*, San Diego: Center for U.S.–Mexican Studies, University of California, 22–6.

Klemenčič, V. and Gosar, A. (1987) 'Grenzüberschreitende Raumwirksame leitbilder dargestellt an Beispielen der Grenzräume Sloweniens in Jugoslawien', *Revue Géographique de l'Est*, 27(1/2): 27–38.

Kliot, N. (1996) 'Turning desert to bloom: Israeli–Jordanian peace proposals for the Jordan rift valley', *Journal of Borderlands Studies*, 11(1): 1–24.

Kliot, N. (1997) 'Regional development for the peace era – Jordanian and Israeli Perspectives', in G. Blake, L. Chia, C. Grundy-War, M. Pratt, and C. Schofield (eds) *International Boundaries and Environmental Security: Frameworks for Regional Cooperation*, Amsterdam: Kluwer, 279–90.

Kliot, N. and Mansfeld, Y. (1994) 'The dual landscape of a partitioned city: Nicosia', in W.A. Gallusser (ed.) *Political Boundaries and Coexistence*, Bern: Peter Lang, 151–61.

Kliot, N. and Mansfeld, Y. (1997) 'The political landscape of partition: The case of Cyprus', *Political Geography*, 16(6): 495–521.

Knight, D.B. (1982) 'Identity and territory: Geographical perspectives on nationalism and regionalism', *Annals of the Assocation of American Geographers*, 72: 514–31.

Knight, D.B. (1994) 'People together, yet apart: Rethinking territory, sovereignty, and identities', in G.J. Demko and W.B. Wood (eds) *Reordering the World: Geopolitical Perspectives on the Twenty-first Century*, Boulder, CO: Westview Press, 71–86.

Koenig, H. (1981) 'The two Berlins', *Travel Holiday*, 156(4): 58–63, 79–80.

Kolossov, V. and O'Laughlin, J. (1998) 'New borders for new world orders: Territorialities at the *fin-de-siecle*', *GeoJournal*, 44(3): 259–73.

Korona, K. (1995) 'The border should help cooperation, not block it', *International Affairs*, 6: 92–4.

Korten, D.C. (1981) 'Management of social transformation', *Public Administration Review*, 41: 609–18.

Koter, M. (1994) 'Transborder "Euroregions" around Polish border zones as an example of a new form of political coexistence', in W.A. Gallusser (ed.) *Political Boundaries and Coexistence*, Bern: Peter Lang, 77–87.

Kotkin, S. and Wolff, D. (eds) (1995) *Rediscovering Russia in Asia: Serbia and the Russian Far East,* Armonk, NY: M.E. Sharp.

Kovács, Z. (1989) 'Border changes and ther effect on the structure of Hungarian society', *Political Geography Quarterly*, 8(1): 79–86.

Krakover, S. (1985) 'Development of tourism resort areas in arid regions', in Y. Gradus (ed.) *Desert Development: Man and Technology in Sparselands*, Dordrecht: D. Reidel Publishing, 271–84.

Krätke, S. (1998) 'Problems of cross-border regional integration: The case of the German–Polish border area', *European Urban and Regional Studies*, 5(3): 249–62.

Kreck, L.A. (1985) 'The effect of the across-the-border commerce of Canadian tourists on the city of Spokane', *Journal of Travel Research*, 24(1): 27–31.

Kresl, P.K. (1993) 'The impact of free trade on Canadian–American border cities', *Canadian–American Public Policy*, 16: 1–44.

Kristoff, L.K.D. (1959) 'The nature of frontiers and boundaries', *Annals of the Association of American Geographers*, 49: 269–82.

Kustanowitz, S. (1994) 'Peace agreement produces new border crossing', *Travel Weekly*, 53(63): 7.

La Ganga, M.L. (1995) 'Pit stop on Nevada border now a hot spot', *Los Angeles Times*, 11 April: A1.

Laar, M., Birkavs, V., and Slezevicius, A. (1996) 'Free trade agreement between the Republic of Estonia, the Republic of Latvia, and the Republic of Lithuania', *Russian and East European Finance and Trade*, 32(5): 39–48.

Lampmann, J. (1997) 'Argentina side trips open the door to wonder and adventure', *Christian Science Monitor*, 17 July: 13.

Langford, D.L. (1998) 'It's party time in St Maarten', *The Toronto Sun*, 12 April: T14.

Lapidoth, R. (1986) 'The Taba controversy', *The Jerusalem Quarterly*, 37: 29–39.

Lee, Y.L. (1980) *The Razor's Edge: Boundaries and Boundary Disputes in Southeast Asia*, Singapore: Institute of Southeast Asian Studies.

Lefebvre, H. (1991) *The Production of Space*, Oxford: Basil Blackwell.

Leifer, M. (1962) 'Cambodia and her neighbours', *Pacific Affairs*, 34(4): 361–74.

Leimgruber, W. (1980) 'Die Grenze als Forschungsobjekt der Geographie', *Regio Basiliensis*, 21: 67–78.

Leimgruber, W. (1981) 'Political boundaries as a factor in regional integration: Examples from Basle and Ticino', *Regio Basiliensis*, 22: 192–201.

Leimgruber, W. (1988) 'Border trade: The boundary as an incentive and an obstacle to shopping trips', *Nordia*, 22(1): 53–60.

Leimgruber, W. (1989) 'The perception of boundaries: Barriers or invitation to interaction?', *Regio Basiliensis*, 30: 49–59.

Leimgruber, W. (1991) 'Boundary, values and identity: The Swiss–Italian transborder region', in D. Rumley and J.V. Minghi (eds) *The Geography of Border Landscapes*, London: Routledge, 43–62.

Leimgruber, W. (1998) 'Defying political boundaries: Transborder tourism in a regional context', *Visions in Leisure and Business*, 17(3): 8–29.

Leiper, N. (1989) 'Tourism and gambling', *GeoJournal*, 19(3): 269–75.

Lew, A.A. (1996) 'Tourism management on American Indian lands in the USA', *Tourism Management*, 17: 355–65.

Lew, A.A. and Van Otten, G.A. (eds) (1998) *Tourism and Gaming on American Indian Lands*, New York: Cognizant.

Lewis, K. (1990) 'Buying across the border', *Canadian Consumer*, 20(3): 9–14.

Lewis, P.F. (1979) 'Axioms for reading the landscape: Some guides to the American scene', in D.W. Meinig (ed.) *The Interpretation of Ordinary Landscapes: Geographical Essays*, New York: Oxford University Press, 11–31.

Liddle, R.W. and Mallarangeng, R. (1997) 'Indonesia in 1996: Pressures from above and below', *Asian Survey*, 37: 167–74.

Liechtenstein National Tourist Office (1997) *Principality of Liechtenstein: Tourist Guide 1998*, Vaduz: National Tourist Office.

Lieff, B.C. and Lusk, G. (1990) 'Transfrontier cooperation between Canada and the USA: Waterton-Glacier International Peace Park', in J. Thorsell (ed.) *Parks on the Borderline: Experience in Transfrontier Conservation*, Gland: IUCN, 39–49.

Light, D. (2000) 'Gazing on communism: Heritage tourism and post-communist identities in Germany, Hungary and Romania', *Tourism Geographies*, 2(2): 157–76.

Lin, V.L. and Loeb, P.D. (1977) 'Tourism and crime in Mexico: Some comments', *Social Science Quarterly*, 58(1): 164–7.

Lintner, B. (1991a) 'Forgotten frontiers: Peace brings investors to notorious border region', *Far Eastern Economic Review*, 16 May: 23–4.

Lintner, B. (1991b) 'Upstaging Macau: Casino at centre of border development plan?', *Far Eastern Economic Review*, 16 May: 24.

Livingston, D.N., Keane, M.C., and Boal, F.W. (1998) 'Space for religion: A Belfast case study', *Political Geography*, 17: 145–70.

Llivia Municipal Museum Trust (1986) *Vila de Llivia*, Llivia, Spain: Municipal Museum Trust.

Lloyd, B. and Swift, T. (1993) 'Adventures north', *The Leader*, 23(5): 6.

Lloydminster Tourism and Convention Authority (n.d.) *Lloydminster: Make a Break for the Border*, Lloydminster: Tourism and Convention Authority.

Lockhart, D.G. (1993) 'Tourism and politics: The example of Cyprus', in D.G. Lockhart, D. Drakakis-Smith, and J. Schembri (eds) *The Development Process in Small Island States*, London: Routledge, 228–46.

Lockhart, D.G. (1997a) 'Islands and tourism: An overview', in D.G. Lockhart and D. Drakakis-Smith (eds) *Island Tourism: Trends and Prospects*, London: Pinter, 3–20.

Lockhart, D.G. (1997b) 'Tourism in Malta and Cyprus', in D.G. Lockhart and D. Drakakis-Smith (eds) *Island Tourism: Trends and Prospects*, London: Pinter, 152–78.

Lockhart, D.G. and Ashton, S. (1990) 'Tourism to Northern Cyprus', *Geography*, 75(2): 163–7.

London Free Press (1992) 'YES M!CH!GAN', *London Free Press*, 7 March: F10.

Long, V.H. (1993) 'Techniques for socially sustainable tourism development: Lessons from Mexico', in J.G. Nelson, R.W. Butler, and G. Wall (eds) *Tourism and Sustainable Development: Monitoring, Planning, Managing*, Waterloo, ON: University of Waterloo, Department of Geography, 201–18.

Lopez, B. (1989) *Crossing Open Ground*, New York: Vintage Books.

Lösch, A. (1954) *The Economics of Location*, New Haven, CT: Yale University Press.

Love, T. (1998) 'Uganda: Beautiful but AIDS stricken', *Tennessee Alumnus*, 78(2): 40–1.

Low, L. and Heng, T.M. (1997) 'Singapore: Development of gateway tourism', in F.M. Go and C.L. Jenkins (eds) *Tourism and Economic Development in Asia and Australasia*, London: Cassell, 236–54.

Lowenthal, D. (1961) 'Geography, experience, and imagination: Towards a geographical epistemology', *Annals of the Association of American Geographers*, 51: 241–61.

Lucas, R.C. (1964) 'Wilderness perception and use: The example of the Boundary Waters canoe area', *Natural Resources Journal*, 3(3): 394–411.

Lundberg, D.E. and Lundberg, C.B. (1993) *International Travel and Tourism*, Second Edition, New York: Wiley.

Lyons, J. (1991) 'Border merchants', *Forbes*, 19 August: 56–7.

Macau Government Tourist Office (1994) *Bestway to Macau*, Macau: Government Tourist Office.

Mackay, J.R. (1958) 'The interactance hypothesis and boundaries in Canada: A preliminary study', *Canadian Geographer*, 11: 1–8.

Madden, K. (1995) 'The best of the Berkshires', *Travel & Leisure*, 25(8): 76–83.

Maier, J. and Weber, J. (1979) 'Tourism and leisure behaviour subject to the spatial influence of a national frontier: The example of Northeast Bavaria', in G. Gruber, H. Lamping, W. Lutz, J. Matznetter, and K. Vorlaufer (eds) *Tourism and Borders: Proceedings of the Meeting of the IGU Working Group – Geography of Tourism and Recreation*, Frankfurt: Institut für Wirtschafts- und Sozialgeographie der Johann Wolfgang Goethe Universität, 111–27.

Maier, M.A. (1992) 'Border jumpers', *Hispanic*, 5(9): 70.

Maillat, D. (1990) 'Transborder regions between members of the EC and non-member countries', *Built Environment*, 16(1): 38–51.

Mann, R. (1987) 'The world's oldest republic: San Marino', *Los Angeles Times*, 20 December: 1, 10.

Mansfeld, Y. (1996) 'Wars, tourism and the "Middle East" factor', in A. Pizam and Y. Mansfeld (eds) *Tourism, Crime and International Security Issues*, Chichester: Wiley, 265–78.

Mansfeld, Y. and Kliot, N. (1996) 'The tourism industry in the partitioned island of Cyprus', in A. Pizam and Y. Mansfeld (eds) *Tourism, Crime and International Security Issues*, Chichester: Wiley, 187–202.

Marić, R. (1988) 'Osnovna prostorno-fizionomska i sadržinska obeležja pograničnog prostora SR Srbije van SAP', *Teorija i Praksa Turizma*, 1: 26–31.

Marks, I.M. (1987) *Fears, Phobias, and Rituals: Panic, Anxiety, and Their Disorders*, New York: Oxford University Press.

Martin, L. (1938) 'The second Wisconsin–Michigan boundary case in the supreme court of the United States, 1930–1936', *Annals of the Association of American Geographers*, 28: 77–126.

Martinez, O.J. (1988) *Troublesome Border*, Tucson: University of Arizona Press.

Martinez, O.J. (1994) 'The dynamics of border interaction: New approaches to border analysis', in C.H. Schofield (ed.) *World Boundaries, Vol. 1, Global Boundaries*, London: Routledge, 1–15.

Martinez, O.J. (1996) 'Introduction', in O.J. Martinez (ed.) *U.S.–Mexico Borderlands: Historical and Contemporary Perspectives*, Wilmington, DE: Scholarly Resources, xiii–xix.

Maryland Office of Tourism Development (1989) *Maryland Travel and Outdoor Guide*, Baltimore, MD: Maryland Office of Tourism Development.

Mashinsky, V. (1996) 'Border tax on the rich', *Izvestia*, 18 December: 2.

Maslow, A. (1954) *Motivation and Personality*, New York: Harper and Row.

Mathieson, A. and Wall, G. (1982) *Tourism: Economic, Physical and Social Impacts*, London: Longman.

Matley, I.M. (1976) *The Geography of International Tourism*, Washington, DC: Association of American Geographers.

Matley, I.M. (1977) 'Physical and cultural factors influencing the location of tourism', in E.M. Kelly (ed.) *Domestic and International Tourism*, Wellesley, MA: The Institute of Certified Travel Agents, 16–25.

Matznetter, J. (1979) 'Border and tourism: Fundamental relations', in G. Gruber, H. Lamping, W. Lutz, J. Matznetter, and K. Vorlaufer (eds) *Tourism and Borders: Proceedings of the Meeting of the IGU Working Group – Geography of Tourism and*

Recreation, Frankfurt: Institut für Wirtschafts- und Sozialgeographie der Johann Wolfgang Goethe Universität, 61–73.

Mayes, H.G. (1992) 'The International Peace Garden: A border of flowers', *The Beaver*, 72(4): 45–51.

McAllister, B. (1996) 'Canadians find shelter in U.S. border enclave', *Washington Post*, 14 May.

McAllister, H.E. (1961) 'The border tax problem in Washington', *National Tax Journal*, 14(4): 362–74.

McGreevy, P. (1988) 'The end of America: The beginning of Canada', *Canadian Geographer*, 32(4): 307–18.

McGreevy, P. (1991) *The Wall of Mirrors: Nationalism and Perceptions of the Border at Niagara Falls*. Borderlands Monograph Series, No. 5, Orono, ME: University of Maine, The Canadian–American Center, 1–18.

McKenna, B. (1997a) 'Review of Canada–U.S. border dispute months away', *The Globe and Mail* (Toronto), 8 October: A17.

McKenna, B. (1997b) 'U.S. border law attacked', *The Globe and Mail* (Toronto), 15 October: A1, A16.

McKenna, B. (1998) 'Crossing U.S. border still easy – for now', *The Globe and Mail* (Toronto), 29 September: A5

McKinsey, L.S. and Konrad, V.A. (1989) *Borderlands Reflections: The United States and Canada,* Orono, ME: University of Maine, Borderlands Project.

McMillan, C. (1993) *Building Blocks or Trade Blocs: NAFTA, Japan and the New World Order*, Ottawa: Canada–Japan Trade Council.

McNeil, D.G. (1997) 'Out of Pretoria by luxury train', *New York Times*, 12 October: 14.

McWhirter, W. (1992) 'A monster spending spree', *Time*, 23 March: 7.

Medvedev, S. (1999) 'Across the line: Borders in post-Westphalian landscapes', in H. Eskelinen, I. Liikanen, and J. Oksa (eds) *Curtains of Iron and Gold: Reconstructing Borders and Scales of Interaction*, Aldershot: Ashgate, 43–56.

Meldman, M. (1995) 'Four faces of Mexico: Along the U.S. border, boundless bargains', *Washington Post*, 24 September: E1.

Merrett, C. (1984) 'The significance of the political boundary in the apartheid state, with particular reference to Transkei', *South African Geographical Journal*, 66(1): 79–93.

Merrett, C. (1991) *Crossing the Border: The Canada–United States Boundary*. Borderlands Monograph Series, No. 5, Orono, ME: University of Maine, The Canadian-American Center, 19–54.

Mihalič, T. (1996) 'Tourism and warfare: The case of Slovenia', in A. Pizam and Y. Mansfeld (eds) *Tourism, Crime and International Security Issues*, Chichester: Wiley, 231–46.

Mihalik, B.J. (1992) 'Tourism impacts related to EC 92: A look ahead', *Journal of Travel Research*, 31(2): 27–33.

Mikesell, J.L. (1970) 'Central cities and sales tax rate differentials: The border city problem', *National Tax Journal*, 23(2): 206–13.

Mikesell, J.L. (1971) 'Sales taxation and the border county problem', *Quarterly Review of Economics and Business*, 11(1): 23–9.

Mikus, W. (1986) 'Grenzüberschreitende verflechtungen im tertiären Sektor zwischen USA und Mexiko: Das Beispiel Kaliforniens', *Geographica Helvetica*, 36: 207–17.

Mikus, W. (1994) 'Research methods in border studies: Results for Latin America', in W.A. Gallusser (ed.) *Political Boundaries and Coexistence*, Bern: Peter Lang, 441–9.

Milne, S. (1992) 'Tourism and development in South Pacific microstates', *Annals of Tourism Research*, 19: 191–212.

Milner, M. and Brummer, A. (1994) 'North and south put heads together on tourism', *Guardian*, 4 February.

Minghi, J.V. (1963a) 'Boundary studies in political geography', *Annals of the Association of American Geographers*, 53: 407–28.

Minghi, J.V. (1963b) 'Television preference and nationality in a boundary region', *Sociological Inquiry*, 33(2): 165–79.

Minghi, J.V. (1981) 'The Franco–Italian borderland: Sovereignty change and contemporary developments in the Alpes-Maritimes', *Regio Basiliensis*, 22: 232–46.

Minghi, J.V. (1991) 'From conflict to harmony in border landscapes', in D. Rumley and J.V. Minghi (eds) *The Geography of Border Landscapes*, London: Routledge, 15–30.

Minghi, J.V. (1994a) 'European borderlands: International harmony, landscape change and new conflict', in C. Grundy-Warr (ed.) *World Boundaries, Vol. 3, Eurasia*, London: Routledge, 89–98.

Minghi, J.V. (1994b) 'The impact of Slovenian independence on the Italo–Slovene borderland: An assessment of the first three years', in W.A. Gallusser (ed.) *Political Boundaries and Coexistence*, Bern: Peter Lang, 88–94.

Minghi, J.V. (1999) 'Borderland "day tourists" from the East: Trieste's transitory shopping fair', *Visions in Leisure and Business*, 17(4): 32–49.

Minghi, J.V. and Rumley, D. (1972) 'Integration and system stress in an international enclave community: Point Roberts, Washington', *B.C. Geographical Series*, 15: 213–29.

Ministère du Tourisme (1993) *Estrie Eastern Townships*, Quebec City: Ministère du Tourisme, Gouvernement du Québec.

Ministry of the Interior (1974) *Instructions Concerning Movement and Stay in Frontier Zone*, Helsinki: Ministry of the Interior, Headquarters of the Frontier Guards.

Mo, J. (1994) 'German lessons for managing the economic cost of Korean reunification', in T.H. Henriksen and K. Lho (eds) *One Korea? Challenges and Prospects for Reunification*, Stanford, CA: Hoover Institution Press, Stanford University, 48–67.

Modler, H. and Boisclair, M. (1995) 'From baubles to bean pots: Shoppers won't be disappointed', *Meetings & Conventions*, 30(3): 104–8.

Moedjanto, G. (1986) *The Concept of Power in Javanese Culture*, Yogyakarta: Gadjah Mada University Press.

Moffett, G.D. (1988a) 'Disputed patch of desert a key to Egyptian–Israeli relations', *Christian Science Monitor*, 8 June: 11.

Moffett, G.D. (1988b) 'Taba verdict: Boost to Israel–Egypt ties?', *Christian Science Monitor*, 28 September: 7–8.

Mok, C. and Dewald, B. (1999) 'Tourism in Hong Kong: After the handover', *Asia Pacific Journal of Tourism Research*, 3(2): 32–40.

Morehouse, B.J. (1996) 'Conflict, space, and resource management at Grand Canyon', *Professional Geographer*, 48(1): 46–57.

Muha, S. (1977) 'Who uses highway welcome centers?', *Journal of Travel Research*, 15(3): 1–4.

Murphy, P.E. (1985) *Tourism: A Community Approach*, London: Methuen.

Naidu, G. (1988) 'ASEAN cooperation in transport', in H. Esmara (ed.) *ASEAN Economic Cooperation: A New Perspective*, Singapore: Chopmen Publishers, 191–204.

Navajo Parks and Recreation Department (n.d.) *Four Corners Monument Navajo Tribal Park*, Window Rock, AZ: Navajo Parks and Recreation Department.

New Brunswick Tourism (1997) *Oh, Say Can You Save!*, Fredericton: New Brunswick Tourism.

Nezavisimaya Gazeta (1994) 'Special conditions for crossing the border are relaxed', *Nezavisimaya Gazeta*, 28 December: 1.

Ngamsom, B. (1998) 'Shopping tourism: A case study of Thailand', in K.S. Chon (ed.) *Proceedings, Tourism and Hotel Industry in Indo-China and Southeast Asia: Development, Marketing, and Sustainability*, Houston, TX: University of Houston, 112–28.

Niagara Parks Commission (1992) *Niagara Falls: The Wonder of the World*, Niagara Falls, ON: Niagara Parks Commission.

Nieves, E. (1996) 'Casino envy gnaws at Falls on U.S. side', *New York Times*, 15 December: 49.

Nin, C.Y. (1994) 'Trade at the Sino–Kazakhstani border: A visit to Korgas and Yining', *China Tourism*, 167: 60–5.

Nin, C.Y. (1998) 'A boundary waterfall', *China Tourism*, 215: 10–15.

North Dakota Parks and Tourism (1992) *Discover the Spirit!*, Bismark: North Dakota Parks and Tourism.

Norton, P. (1989) 'Archaeological rescue and conservation in the north Andean area', in H. Cleere (ed.) *Archaeological Heritage Management in the Modern World*, London: Unwin Hyman, 142–5.

Office du Tourisme (1996) *Reflets, ile de Saint Martin*, Marigot: Office du Tourisme de l'ile de Saint Martin.

O'Dowd, L., Corrigan, J., and Moore, T. (1995) 'Borders, national soveriegnty and European integration: The British–Irish case', *International Journal of Urban and Regional Research*, 19(2): 272–85.

Olsen, D.H. and Timothy, D.J. (1999) 'Tourism 2000: Selling the Millennium', *Tourism Management*, 20: 389–92.

O'Neil, D. (1996) 'St Maarten: Double your pleasure', *Leisure World*, 8(1): 16–18.

Oppermann, M. (1998) 'Tourism space revisited', *Tourism Analysis*, 2: 107–18.

Outside (1994) 'Big Bend National Park', *Outside*, 19(7): 63–8.

Paasi, A. (1994) 'The changing representations of the Finnish–Russian boundary', in W.A. Gallusser (ed.) *Political Boundaries and Coexistence*, Bern: Peter Lang, 103–11.

Paasi, A. (1996) *Territories, Boundaries and Consciousness: The Changing Geographies of the Finnish–Russian Border*, Chichester: Wiley.

Paasi, A. and Raivo, P.J. (1998) 'Boundaries as barriers and promoters: Constructing the tourist landscapes of Finnish Karelia', *Visions in Leisure and Business*, 17(3): 30–45.

Page, S.J. (1994) 'Perspectives on tourism and peripherality: Review of tourism in the Republic of Ireland', *Progress in Tourism, Recreation and Hospitality Management*, 5: 26–53.

Pagnini, M.P. (1979) 'Friuli-Venetia Julia: A tourist border region', in G. Gruber, H. Lamping, W. Lutz, J. Matznetter, and K. Vorlaufer (eds) *Tourism and Borders: Proceedings of the Meeting of the IGU Working Group – Geography of Tourism and Recreation*, Frankfurt: Institut für Wirschafts- und Sozialgeographie der Johann Wolfgang Goethe Universität, 205–13.

Palomäki, M. (1994) 'Transborder cooperation over Quarken Strait between Finland and Sweden', in W.A. Gallusser (ed.) *Political Boundaries and Coexistence*, Bern: Peter Lang, 238–46.

Panic-Kombol, T. (1996) 'The cultural heritage of Croatian cities as a tourism potential', *World Leisure and Recreation*, 38(1): 21–5.

Parent, L. (1990) 'Tex-Mex Park: Making Mexico's Sierra del Carmen a sister park to Big Bend', *National Parks*, 64(7): 30–6.

Parfit, M. (1996) 'Tijuana and the border: Magnet of opportunity', *National Geographic*, 190(2): 94–107.

Patrick, J.M. and Renforth, W. (1996) 'The effects of the peso devaluation on cross-border retailing', *Journal of Borderlands Studies*, 11(1): 25–41.

Pavlakovic, V.K. and Kim, H.H. (1990) 'Outshopping by maquila employees: Implications for Arizona's border communities', *Arizona Review*, Spring: 9–16.

Payne, T. (1987) 'Economic issues', in C. Clarke and T. Payne (eds) *Politics, Security and Development in Small States*, London: Allen & Unwin, 50–62.

Pearce, D.G. (1992) 'Tourism and the European Regional Development Fund: The first fourteen years', *Journal of Travel Research*, 30(3): 44–51.

Pearce, D.G. (1995) 'CER, trans-Tasman tourism and a single aviation market', *Tourism Management*, 16: 111–20.

Pearce, D.G. (1998) 'Tourist districts in Paris: Structure and functions', *Tourism Management*, 19: 49–65.

Pearce, D.G. (1999a) 'Introduction: Issues and approaches', in D.G. Pearce and R.W. Butler (eds) *Contemporary Issues in Tourism Development*, London: Routledge, 1–12.

Pearce, D.G. (1999b) 'Tourism in Paris: Studies at the microscale', *Annals of Tourism Research*, 26: 77–97.

Pearce, P.L., Moscardo, G. and Ross, G.F. (1996) *Tourism Community Relationships*. Oxford: Pergamon.

Pearcy, G.E. (1965) 'Boundary functions', *Journal of Geography*, 64(8): 346–9.

Pedreschi, L. (1957) 'L'exclave Italiano in terra Svizzera di Campione d'Italia', *Revista Geografica Italiana*, 64: 23–40.

Perry, M. (1991) 'The Singapore Growth Triangle: State, capital, and labour at a new frontier in the world economy', *Singapore Journal of Tropical Geography*, 12(2): 138–51.

Pertman, A. (1995) 'California trolley enhances border appeal', *Boston Globe*, 14 August: 3.

Pizam, A. (1982) 'Tourism and crime: Is there a relationship?', *Journal of Travel Research*, 20(3): 7–10.

Pollack, A. (1996a) 'At the DMZ, another invasion: Tourists', *New York Times*, 10 April: A10.

Pollack, A. (1996b) 'Behind North Korea's barbed wire: Capitalism', *New York Times*, 15 September: A9.

Pommersheim, F. (1989) 'The crucible of sovereignty: Analyzing issues of tribal jurisdiction', *Arizona Law Review*, 31(2): 329–63.

Porter, J. (1997) 'Macau 1999', *Current History*, 96: 282–6.

Prescott, J.R.V. (1987) *Political Frontiers and Boundaries*, London: Allen and Unwin.

Pretes, M. (1995) 'Postmodern tourism: The Santa Claus industry', *Annals of Tourism Research*, 22: 1–15.

Prock, J. (1983) 'The peso devaluations and their effect on Texas border economies', *Inter-American Economic Affairs*, 37(3): 83–92.

Puente, M. (1996) 'So close, yet so far: San Diego, Tijuana bridging gap', in O.J. Martinez (ed.) *U.S.–Mexico Borderlands: Historical and Contemporary Perspectives*, Wilmington, Delaware: Scholarly Resources, 249–55.

Raafat, W. (1983) 'The Taba case between Egypt and Israel', *Revue Egyptienne de Droit International*, 39: 1–22.

Randall, J. and Conrad, H.W. (eds) (1995) *NAFTA in Transition*, Calgary: University of Calgary Press.

Randerson, M. (1994) 'There's a casino just across the border', *Houston Post*, 30 January: F1.

Ratzel, F. (1892) 'Die politischen Grenzen', *Mitteilungen der Geographischen Gesellschaft für Thüringen zu Jena*, 11: 69–73.

Ratzel, F. (1896) 'The territorial growth of states', *Scottish Geographical Magazine*, 12: 351–61.

Reaves, J.A. (1989) 'Border menacing despite changes', *Chicago Tribune*, 12 November: 4.

Reichart, T. (1988) 'Socio-economic difficulties in developing tourism in small Alpine countries: The case of Andorra', *Tourism Recreation Research*, 13(1): 27–32.

Reitsma, H.J. (1971) 'Crop and livestock production in the vicinity of the United States–Canada border', *The Professional Geographer*, 23(3): 216–23.

Reitsma, H.J. (1972) 'Areal differentiation along the United States–Canada border', *Tijdschrift voor Economische and Sociale Geografie*, 63: 2–10.

Renard, J.P. (1994) 'L'Aménagement du territoire en France et les frontières', *Hommes et Terres du Nord*, 2(3): 95–102.

Rennicke, J. (1995) 'Rafting the Rio', *National Geographic Traveler*, 12(2): 106–15.

Reynolds, C. (1997) 'U.S. ban aside, Americans visit Cuba easily', *Los Angeles Times*, 15 June: F1–2.

Reynolds, D.R. and McNulty, M.L. (1968) 'On the analysis of political boundaries as barriers: A perceptual approach', *East Lakes Geographer*, 4: 21–38.

Reza, H.G. (1995) 'U.S. eyes open border with Canada', *The Toronto Star*, 27 August: 3.

Richard, W.E. (1993) 'International planning for tourism', *Annals of Tourism Research*, 20: 601–4.

Richard, W.E. (1995) 'Segmenting cross-border tourism with a focus on the shopping component', in B.D. Middlekauff (ed.) *Proceedings of the Annual Meeting of the New England-St Lawrence Valley Geographical Society*, Burlington, VT: New England-St Lawrence Valley Geographical Society, 183–97.

Richardson, H.L. (1993) 'NAFTA means economic growth', *Transportation and Distribution*, 34(11): 42–8.

Richter, L.K. (1992) 'Political instability and tourism in the Third World', in D. Harrison (ed.) *Tourism and the Less Developed Countries*, London: Belhaven, 35–46.

Rimmer, P.J. (1994) 'Regional economic integration in Pacific Asia', *Environment and Planning A*, 26: 1731–59.

Rinehart, D. (1992) 'Canadians making fewer trips south', *London Free Press*, 15 August: B1.

Ringer, G. (1998) 'Introduction', in G. Ringer (ed.) *Destinations: Cultural Landscapes of Tourism*, London: Routledge, 1–13.

Rinschede, G. (1977) 'Andorra: vom abgeschlossenen Hochgebirgsstaat zum internationalen Touristenzentrum', *Erdkunde*, 31: 307–14.

Ritchie, K.D. (1993) 'Spatial analysis of cross-border shopping in Southern Ontario', unpublished B.A. honor's thesis, Department of Geography, University of Waterloo.

Ritter, G. and Hajdu, J.G. (1989) 'The east–west German boundary', *Geographical Review*, 79(3): 326–44.

Roberts, G.K. (1991) 'Emigrants in their own country: German reunification and its political consequences', *Parliamentary Affairs*, 44: 373–88.

Roberts, J. (1995) 'Israel and Jordan: Bridges over the borderlands', *Boundary and Security Bulletin*, 2(4): 81–4.

Robinson, G. and Mogendorff, D. (1993) 'The European tourism industry: Ready for a single market?', *International Journal of Hospitality Management*, 12: 21–31.

Robinson, G.W.S. (1959) 'Exclaves', *Annals of the Association of American Geographers*, 49: 283–95.

Rodríguez, M. and Portales, J. (1994) 'Tourism and NAFTA: Towards a regional tourism policy', *Tourism Management*, 15: 319–22.

Roehl, W.S. (1995) 'The June 4, 1989, Tiananmen Square incident and Chinese tourism', in A.A. Lew and L. Yu (eds) *Tourism in China: Geographic, Political, and Economic Perpsectives*, Boulder, CO: Westview Press, 19–39.

Rojek, C. (1998) 'Cybertourism and the phantasmagoria of place', in G. Ringer (ed.) *Destinations: Cultural Landscapes of Tourism*, London: Routledge, 33–48.

Rogerson, C.M. (1990) 'Sun International: The making of a South African tourism multinational', *GeoJournal*, 22(3): 345–54.

Rossides, N. (1995) 'The conservation of the cultural heritage in Cyprus: A planner's perspective', *Regional Development Dialogue*, 16(1): 110–25.

Royle, S.A. (1997) 'Tourism in the South Atlantic islands', in D.G. Lockhart and D. Drakakis-Smith (eds) *Island Tourism: Trends and Prospects*, London: Pinter, 323–44.

Rozman, G. (1995) 'Spontaneity and direction along the Russo-Chinese border', in S. Kotkin and D. Wolff (eds) *Rediscovering Russia in Asia: Siberia and the Russian Far East*, Armonk, NY: M.E. Sharp, 275–89.

Rumley, D. and Minghi, J.V. (1991) 'Introduction: The border landscape concept', in D. Rumley and J.V. Minghi (eds) *The Geography of Border Landscapes*, London: Routledge, 1–14.

Ruppert, K. (1979) 'Funktionale Verflechtungen im deutsch-österreichischen grenzraum', in G. Gruber, H. Lamping, W. Lutz, J. Matznetter, and K. Vorlaufer (eds) *Tourism and Borders: Proceedings of the Meeting of the IGU Working Group – Geography of Tourism and Recreation*, Frankfurt: Institut für Wirtschafts- und Sozialgeographie der Johann Wolfgang Goethe Universität, 95–110.

Ryan, C. (1991) *Recreational Tourism: A Social Science Perspective*, New York: Routledge.

Ryan, C. (1993) 'Crime, violence, terrorism and tourism: An accidental or intrinsic relationship?', *Tourism Management*, 14: 173–83.

Ryden, K.C. (1993) *Mapping the Invisible Landscape: Folklore, Writing, and the Sense of Place*, Iowa City: University of Iowa Press.

Sack, R.D. (1986) *Human Territoriality: Its Theory and History*, Cambridge: Cambridge University Press.

Saint-Germain, M.A. (1995) 'Problems and opportunities for cooperation among public managers on the U.S.–Mexico border', *American Review of Public Administration*, 25(2): 93–117.

Salomon, J.N. (1992) 'Le complexe touristico-industriel d'Iguaçu-Itaipu (Argentine-Brésil-Paraguay)', *Cahiers d'Outre-Mer*, 45: 5–20.
San Diego Convention and Visitors Bureau (1999) *Official Visitors Planning Guide*, San Diego: Convention and Visitors Bureau.
Sánchez, M.L. (ed.) (1994) *A Shared Experience: The History, Architecture and Historic Designations of the Lower Rio Grande Heritage Corridor*, Austin, TX: Los Caminos del Rio Heritage Project, Texas Historical Commission.
Sándor, J. (1990) 'A nyugati határszél idegenforgalma a vonzás és a keresletkinálat tükrében', *Idegenforgalmi Közlemények*, 1: 25–34.
Sanguin, A.L. (1974) 'La frontière Québec-Maine: quelques aspects limologiques et socio-économiques', *Cahiers de Géographie de Québec*, 18(43): 159–85.
Sanguin, A.L. (1991) 'L'Andorre ou la quintessence d'une economie transfrontaliere', *Revue Geographique des Pyrenees et du Sud-Ouest*, 62(2): 169–86.
Scalapino, R.A. (1992) 'Northeast Asia – prospects for cooperation', *The Pacific Review*, 5(2): 101–11.
Scanian, D. (1991a) 'Hard times on the line', *The Ottawa Citizen*, 18 May: B2.
Scanian, D. (1991b) 'The recession: Canadian dollars helping Massena ride it out', *The Ottawa Citizen*, 18 May: B1.
Schneider, H. (1998) 'Canadian dollar hits record low', *Washington Post*, 23 January.
Schrambling, R. (1991) 'St Martin/Sint Maarten', *Islands*, 11(1): 100–10.
Schwartz, F.D. (1997) 'Niagara Falls: For two hundred years it's been attracting tourists – and tourist traps', *American Heritage*, September: 22–6.
Scott, J. (1995) 'Sexual and national boundaries in tourism', *Annals of Tourism Research*, 22: 385–403.
Scott, J.W. (1989) 'Transborder cooperation, regional initiatives, and sovereignty conflicts in western Europe: The case of the Upper Rhine Valley', *Publius: The Journal of Federalism*, 19: 139–56.
Scott, J.W. (1993) 'The institutionalization of transboundary cooperation in Europe: Recent development on the Dutch–German border', *Journal of Borderlands Studies*, 8(1): 39–66.
Scott, J.W. (1998) 'Planning cooperation and transboundary regionalism: Implementing policies for European border regions in the German–Polish context', *Environment and Planning C*, 16: 605–24.
Seekings, J. (1993) 'Gibraltar: Developing tourism in a political impasse', *Tourism Management*, 14: 61–7.
Seibert, T. (1990) 'What's doing in Berlin', *New York Times*, 22 July: 10.
Sevodnya (1994) 'Russia closes its borders with Azerbaijan and Georgia', *Sevodnya*, 21 December: 1.
Simmons, T. and Turbeville III, D.E. (1984) 'Blaine, Washington: Tijuana of the north?', *B.C. Geographical Series*, 41: 47–57.
Singh, L.P. (1962) 'The Thai–Cambodian temple dispute', *Asian Survey*, 2(8): 23–6.
Sloan, G. (1998) 'Now's a great time to drop a dime in Canada', *USA Today*, 14 August.
Slowe, P.M. (1991) 'The geography of borderlands: The case of the Quebec–US borderlands', *Geographical Journal*, 157(2): 191–8.
Slowe, P.M. (1994) 'The geography of borderlands: The case of the Quebec–US borderlands', in P.O. Girot (ed.) *World Boundaries, Vol. 4, The Americas*, London: Routledge, 3–17.

Small, J. and Witherick, M. (1995) *A Modern Dictionary of Geography,* Third Edition, London: Edward Arnold.

Smith, F.M. (1994) 'Politics, place and German reunification: A realignment approach', *Political Geography,* 13(3): 228–44.

Smith, G. (1994) 'Implications of the North American Free Trade Agreement for the US tourism industry', *Tourism Management,* 15: 323–6.

Smith, G. and Pizam, A. (1998) 'NAFTA and tourism development policy in North America', in E. Laws, B. Faulkner, and G. Moscardo (eds) *Embracing and Managing Change in Tourism: International Case Studies,* London: Routledge, 17–28.

Smith, G. and Hinch, T.D. (1996) 'Canadian casinos as tourist attractions: Chasing the pot of gold', *Journal of Travel Research,* 34(3): 37–45.

Smith, G. and Malkin, E. (1997) 'The border', *BusinessWeek,* 12 May: 64–74.

Smith, S.L.J. (1984) 'A method for estimating the distance equivalence of international boundaries', *Journal of Travel Research,* 22(3): 37–9.

Smith, V.L. (2000) 'Space tourism: The 21st century "frontier"', *Tourism Recreation Research,* 25(3): 5–15.

Sobol, J. (1992) 'Life along the line', *Canadian Geographic,* 112: 46–56.

Sommers, B.J. and Timothy, D.J. (1999) 'Economic development, tourism, and urbanization in the emerging markets of Northeast Asia', in G.P. Chapman, A.K. Dutt, and R.W. Bradnock (eds) *Urban Growth and Development in Asia: Making the Cities,* Aldershot, UK: Ashgate, 111–31.

Sommers, L.M. and Lounsbury, J.F. (1991) 'Border boom towns of Nevada', *Focus,* 41(4): 12–18.

Sonderegger, J. (1996) 'Border war', *St Louis Post-Dispatch,* 25 October.

Sönmez, S.F. (1998) 'Tourism, terrorism, and political instability', *Annals of Tourism Research,* 25: 416–56.

Sönmez, S.F. and Apostolopoulos, Y. (2000) 'Conflict resolution through tourism cooperation? The case of the partitioned island-state of Cyprus', *Journal of Travel and Tourism Marketing,* 9(4): 35–48.

South African Tourism Board (1991) *The Splendor of Five African States,* Pretoria: South African Tourism Board.

South African Tourism Board (1999 – accessed) *Tourism.* Accessed 10 December 1999. http://www.southafrica.net/government/tourism.html

Southall, R.J. (1983) *South Africa's Transkei: The Political Economy of an 'Independent' Bantustan,* New York: Monthly Review Press.

St John, R.B. (1994) 'Preah Vihear and the Cambodia–Thailand borderland', *Boundary and Security Bulletin,* 1(4): 64–8.

Stansfield, C. (1996) 'Reservations and gambling: Native Americans and the diffusion of legalized gaming', in R.W. Butler and T. Hinch (eds) *Tourism and Indigenous Peoples,* London: Routledge, 129–47.

Stansfield, C.A. and Rickert, J.E. (1970) 'The Recreational Business District', *Journal of Leisure Research,* 2(4): 213–25.

Statistics Canada (1986–1998) *Touriscope: International Travel,* Ottawa: Statistics Canada.

Steffens, R. (1993) 'Not just another roadside attraction', *National Parks,* 67(1): 26–31.

Steffens, R. (1994) 'Bridging the border: As NAFTA goes into effect, U.S. and Mexican officials must ensure that economics does not overshadow the need to protect parks and other public lands', *National Parks,* 68(7): 36–41.

Stern, E. (1987) 'Competition and location in the gaming industry: The "casino states" of Southern Africa', *Geography*, 72(2): 140–50.

Stevenson, D. (1991) 'Cross-border dispute', *Canadian Consumer*, 21(7): 8–15.

Stewart, W.P., Lue, C., Fesenmaier, D.R., and Anderson, B.S. (1993) 'A comparison between welcome center visitors and general highway auto travelers', *Journal of Travel Research*, 31(3): 40–6.

Stinson, M. and Bourette, S. (1998) 'Dollar sinks to lowest ever: Americans flock to Canadian border towns to wine and dine at a discount', *The Globe and Mail* (Toronto), 39 January: 1.

Stoddard, E.R. (1987) *Maquila Assembly Plants in Northern Mexico*, El Paso: Texas Western Press.

Storey, R. (1992) *Hong Kong, Macau & Canton*, Hawthorn, Australia: Lonely Planet.

Stryjakiewicz, T. (1998) 'The changing role of border zones in the transforming economies of East-Central Europe: The case of Poland', *GeoJournal*, 44(3): 203–13.

Sturken, B. (1986) 'French isles, airlines differ on visa rules for Americans', *Travel Weekly*, 45: 1.

Successful Meetings (1992) 'St Maarten/St Martin', *Successful Meetings*, 41(12): 35–6.

Suomen Matkailuliitto (1993) *Kalottireitti: Rajaton Tunturielämys Pohjoisessa Suomen, Ruotsin ja Norjan Halki*, Helsinki: Suomen Matkailuliitto.

Sverdlik, A. (1994) 'Fast track to bargains in Tijuana', *Atlanta Journal Constitution*, 3 April: K5.

Swanson, E.J. (1992) 'The reservation gaming craze: Casino gambling under the Indian Gaming and Regulatory Act of 1988'. *Hamline Law Review*, 15: 471–96.

Sweedler, A. (1994) 'Conflict and cooperation in border regions: An examination of the Russian–Finnish border', *Journal of Borderlands Studies*, 9(1): 1–13.

Sweet, A.S. and Sandholtz, W. (1997) 'European integration and supranational governance', *Journal of European Public Policy*, 4(3): 297–317.

Szabo, J. (1996) 'Bonanza on the border: A loco buying spree along the Tex-Mex trail', *Travel & Leisure*, 26(11): 58–64.

Taillefer, F. (1991) 'Le paradoxe andorran', *Revue Geographique des Pyrenees et du Sud-Ouest*, 62(2): 117–38.

Tasker, R., Schwartz, A., and Vatikiotis, M. (1994) 'ASEAN: Growing pains', *Far Eastern Economic Review*, 28 July: 22–3.

Tavris, C. and Wade, C. (1995) *Psychology in Perspective*, New York: Harper Collins.

Taylor, G.D. (1994) 'The implications of free trade agreements for tourism in Canada', *Tourism Management*, 15: 315–56.

Tenhiälä, H. (1994) 'Cross-border cooperation: Key to international ties', *International Affairs*, 6: 21–3.

Texarkana Chamber of Commerce (n.d.) *Texarkana's Trail of Two Cities*, Texarkana: Chamber of Commerce.

Teye, V. (1988) 'Coup d'etat and African tourism: A study of Ghana', *Annals of Tourism Research*, 15: 329–56.

Thorpe, G.L. and Olson, S.L. (1997) *Behavior Therapy: Concepts, Procedures, and Applications*, Second Edition, Boston: Allyn and Bacon.

Thorsell, J. and Harrison, J. (1990) 'Parks that promote peace: A global inventory of transfrontier nature reserves', in J. Thorsell (ed.) *Parks on the Borderline: Experiences in Transfrontier Conservation*, Gland: IUCN, 3–21.

Tierney, P.T. (1993) 'The influence of state traveler information centers on tourist length of stay and expenditures', *Journal of Travel Research*, 31(3): 28–32.

Timothy, D.J. (1995a) 'International boundaries: New frontiers for tourism research', *Progress in Tourism and Hospitality Research*, 1(2): 141–52.

Timothy, D.J. (1995b) 'Political boundaries and tourism: Borders as tourist attractions', *Tourism Management*, 16: 525–32.

Timothy, D.J. (1996) 'Small and isolated: The politics of tourism in international exclaves', *Acta Turistica*, 8(2): 99–115.

Timothy, D.J. (1998a) 'Collecting places: Geodetic lines in tourist space', *Journal of Travel and Tourism Marketing*, 7(4): 123–9.

Timothy, D.J. (1998b) 'Cooperative tourism planning in a developing destination', *Journal of Sustainable Tourism*, 6(1): 52–68.

Timothy, D.J. (1998c) ' International boundaries and tourism: Themes and issues', *Visions in Leisure and Business*, 17(3): 3–7.

Timothy, D.J. (1998d) 'Tourism development in a small, isolated community: The case of Northwest Angle, Minnesota', *Small Town*, 29(1): 20–3.

Timothy, D.J. (1999a) 'Cross-border partnership in tourism resource management: International parks along the US–Canada border', *Journal of Sustainable Tourism*, 7(3/4): 182–205.

Timothy, D.J. (1999b) 'Cross-border shopping: Tourism in the Canada–United States borderlands', *Visions in Leisure and Business*, 17(4): 4–18.

Timothy, D.J. (1999c) 'Participatory planning: A view of tourism in Indonesia', *Annals of Tourism Research*, 26: 371–91.

Timothy, D.J. (2000a) 'Borderlands: An unlikely tourist destination?', *Boundary and Security Bulletin*, 8(1): 57–65.

Timothy, D.J. (2000b) 'Tourism and international parks', in R.W. Butler and S.W. Boyd (eds) *Tourism and National Parks: Issues and Implications*, Chichester: Wiley, 263–82.

Timothy, D.J. (2000c) 'Tourism planning in Southeast Asia: Bringing down borders through cooperation', in K.S. Chon (ed.) *Tourism in Southeast Asia: A New Direction*, New York: The Haworth Hospitality Press, 21–35.

Timothy, D.J. (2001) 'Tourism in the borderlands: Competition, complementarity and cross-frontier cooperation', in S. Krakover and Y. Gradus (eds) *Tourism in Frontier Areas*, Baltimore, MD: Lexington Books, 233–58.

Timothy, D.J. and Butler, R.W. (1995) 'Cross-border shopping: A North American perspective', *Annals of Tourism Research*, 22: 16–34.

Timothy, D.J. and Mao, B. (1992) 'Tourism in international exclaves', paper presented at the annual meeting of the Canadian Association of Geographers, Ontario Division, Scarborough, Ontario, 31 October.

Timothy, D.J. and Wall, G. (1995) 'Tourist accommodation in an Asian historic city', *Journal of Tourism Studies*, 6(2): 63–73.

Timothy, D.J. and White, K. (1999) 'Community-based ecotourism development on the periphery of Belize', *Current Issues in Tourism*, 2(2/3): 226–42.

Ting, Z. (1994) 'Hunchun: A trade city that's going places', *China Today*, 43(1): 22–6.

Toops, S.W. (1992) 'Tourism in China and the impact of June 4, 1989', *Focus*, 42(1): 3–7.

Tourism Canada (1992) *The Tourism Intelligence Bulletin*, Ottawa: Industry, Science and Technology Canada.

Touristikinformation Jungholz (1994) *Tourismusverband Jungholz: Der Kleine Urlaubs-Wanderführer*, Jungholz, Austria: Touristikinformation Jungholz.

Travel and Tourism Executive Report (1997) 'Sec. 110 "non" waiver rules for Canada, Mexico surprised travel industry', *Travel and Tourism Executive Report*, 18(7): 1, 5, 8.

Travel Weekly (1991) *Vendôme Guide: St Martin/St Maarten*, Toronto: *Travel Weekly*.

Travel Weekly (1998) 'Windsor Group opens permanent facility in Ontario', *Travel Weekly*, 5 October: 92.

Truehart, C. (1996) 'The chips' fall bouys Niagara', *Washington Post*, 18 May: A14.

Truitt, L.J. (1996) 'Casino gambling in Illinois: Riverboats, revenues, and economic development', *Journal of Travel Research*, 34(3): 89–96.

Tsang, S.W. (1994) 'Tourist boom to gain pace near 1997', *Eastern Express*, 27 May.

Tucker, K. and Sundberg, M. (1988) *International Trade in Services*, London: Routledge.

Turbeville III, D.E. and Bradbury, S.L. (1997) 'Borderlines, border towns: Cultural landscapes of the post-free trade 49th parallel', paper presented at the annual meeting of the Association of American Geographers, Forth Worth, Texas, April.

Turley, S. (1998) 'Hadrian's Wall (UK): Managing the visitor experience at the Roman frontier', in M. Shackley (ed.) *Visitor Management: Case Studies from World Heritage Sites*, Oxford: Butterworth Heinemann, 100–20.

Turner, P. (1994) *South-East Asia*, Hawthorn, Australia: Lonely Planet.

Turner, L. and Ash, J. (1975) *The Golden Hordes: International Tourism and the Pleasure Periphery*, London: Constable.

Tyler, C. (1992) 'Seas of conflict', *Geographical Magazine*, 64(3): 22–6.

Tzoanos, G. (1993) 'Foreword', *International Journal of Hospitality Management*, 12(1): 1–2.

Ul-chul, Y. (1997) 'DPRK reportedly to designate Pidan Island free trade zone', *Hangyore*, 10 June: 1.

Urry, J. (1995) *Consuming Places*. London: Routledge.

USA Today (1998) 'Finns give Santa a park', *USA Today*, 13 November: 2D.

van Buren, A. (1985) 'Dear Abby', *Buffalo News*, 11 September.

van Tighem, K. (1988) 'Common ground: Drills across the border', *Wilderness*, 52: 54–6.

Varady, R.G., Colnic, D., Merideth, R., and Sprouse, T. (1996) 'The U.S.–Mexican border environment cooperation commission: Collected perspectives on the first two years', *Journal of Borderlands Studies*, 11(2): 89–113.

Vardomsky, L. (1992) 'Blagoveshchensk and Heihe engage in across-the-border cooperation', *Far Eastern Affairs*, 2: 81–6.

Varniere-Simon, F. (1991) 'L'evolution contrastee de l'amanagement touristique transfrontaliere en Ardennes: Perspectives Franco-Belges', *Revue Geographique de l'Est*, 31(2): 113–21.

Verwaltungs- und Privat-Bank (1998) *Liechtenstein in Figures*, Vaduz: Verwaltungs- und Privat-Bank.

Vesilind, P.J. (1990) 'Common ground, different dreams: The U.S.–Canada border', *National Geographic*, 177(2): 94–127.

Viken, A. (1995) 'Tourism experiences in the Arctic: The Svalbard Case', in C.M. Hall and M.E. Johnston (eds) *Polar Tourism: Tourism in the Arctic and Antarctic Regions*, Chichester: Wiley, 73–84.

Viken, A., Vostryakov, L., and Davydov, A. (1995) 'Tourism in Northeast Russia', in C.M. Hall and M.E. Johnston (eds) *Polar Tourism: Tourism in the Arctic and Antarctic Regions*, Chichester: Wiley, 101–14.

Vogel, D. (1997) 'Trading up and governing across: Transnational governance and environmental protection', *Journal of European Public Policy*, 4(4): 556–71.

von Böventer, E. (1969) 'Walter Christaller's central places and peripheral areas: The central place theory in retrospect', *Journal of Regional Science*, 9(1): 117–24.

Vorhes, G. (1990) 'Almost south of the border', *The Western Horseman*, 55(12): 34–8.

Vuoristo, K.V. (1981) 'Tourism in Eastern Europe: Development and regional patterns', *Fennia*, 159(1): 237–47.

Wachowiak, H. (1994) *Grenzüberschreitende Zusammenarbeit im Tourismus: Eine Analyse Grenzüberschreitender Massnahmen Entland der Westlichen Staatsgrenze der Bundesrepublik Deutschland zwischen Ems-Dollart und Baden-Nordelsass-Südpfalz*, Trier: Europäisches Tourismus Institut.

Wackermann, G. (1979) 'Projection socio-culturelle du tourisme et isochrones moyens en espace frontalier', in G. Gruber, H. Lamping, W. Lutz, J. Matznetter, and K. Vorlaufer (eds) *Tourism and Borders: Proceedings of the Meeting of the IGU Working Group – Geography of Tourism and Recreation*, Frankfurt: Institut für Wirtschafts- und Sozialgeographie der Johann Wolfgang Goethe Universität, 295–307.

Walker, A.M. (1994) 'Euro Airport Basle-Mulhouse-Freiburg: Strengths and weaknesses of a bi-national airport', in W.A. Gallusser (ed.) *Political Boundaries and Coexistence*, Bern: Peter Lang, 279–83.

Wall Street Journal (1996) 'Russia to tax travelers', *Wall Street Journal*, 24 December: A6.

Walsh, J. (1992) ' Sea of troubles: China's offshore oil grab chills détente with Vietnam and rings wider Asian alarms', *Time*, 27 July: 40–1.

Wanhill, S. (1997) 'Peripheral area tourism: A European perspective', *Progress in Tourism and Hospitality Research*, 3(1): 47–70.

Warszynska, J. and Jackowski, A. (1979) 'Impact of passport facilities in the passenger traffice between Poland and the German Democratic Republic (GDR) on the development of touristic phenomena', in G. Gruber, H. Lamping, W. Lutz, J. Matznetter, and K. Vorlaufer (eds) *Tourism and Borders: Proceedings of the Meeting of the IGU Working Group – Geography of Tourism and Recreation*, Frankfurt: Institut für Wirtschafts- und Sozialgeographie der Johann Wolfgang Goethe Universität, 353.

Wasserman, D. (1996) 'The borderlands mall: Form and function of an imported landscape', *Journal of Borderlands Studies*, 11(2): 69–88.

Waterman, S. (1987) 'Partitioned states', *Political Geography Quarterly*, 6(2): 151–70.

Wayne, E.A. (1988) 'US sets out to calm two of its Mideast friends', *Christian Science Monitor*, 8 August: 3–4.

Weaver, D.B. (1998) 'Peripheries of the periphery: Tourism in Tobago and Barbuda', *Annals of Tourism Research*, 25: 292–313.

Weaver, G.D. (1966) 'Some spatial aspects of Nevada's gambling economy', paper presented at the annual meeting of the Association of American Geographers, Toronto, August.

Weigand, K. (1990) 'Drei Jahrzehnte Einkaufstourismus über die deutsch-dänische Grenze', *Geographische Rundschau*, 42(5): 286–90.

Weingrod, C. (1994) 'Two countries, one wilderness', *National Parks*, 68(1): 26–31.

West, J.P. and James, D.D. (1983) 'Border tourism', in E.R. Stoddard, R.L. Nostrand, and J.P. West (eds) *Borderlands Sourcebook: A Guide to the Literature on Northern Mexico and the American Southwest*, Norman: University of Oklahoma Press, 159–65.

West, R. (1973) 'Border towns: What to do and where to do it', *Texas Monthly*, 1: 62–73, 109.

Western Arctic Tourism Association (n.d.) *The Dempster Highway: A Road Less Traveled, A Land Unspoiled*, Inuvik, NWT: Western Arctic Tourism Association.

Westing, A.H. (1993) 'Building confidence with transfrontier reserves: The global potential', in A.H. Westing (ed.) *Transfrontier Reserves for Peace and Nature: A Contribution to Human Society*, Nairobi: UNEP, 1–15.

White, M. (1999) 'Entrepreneurs study space tourism with "reality" 15–20 years in future', *Sentinal Tribune*, 23 September: 10.

Wilkinson, P.F. (1987) 'Tourism in small island nations: A fragile dependence', *Leisure Studies*, 6(2): 127–46.

Wilkinson, P.F. (1989) 'Strategies for tourism in island microstates', *Annals of Tourism Research*, 16: 153–77.

Williams, A.M. and Shaw, G. (1998) 'Tourism and economic development', in D. Pinder (ed.) *The New Europe: Economy, Society and Environment*, Chichester: Wiley, 177–201.

Williams, A.M. and Shaw, G. (2000) 'Guest editorial: Tourism geography in a changing world', *Tourism Geographies*, 2(3): 239–40.

Wilson, J. (2000) 'Postcards from the moon: A lunar vacation isn't as far-out an adventure as you think', *Popular Mechanics*, June: 97–9.

Wilson, J.R. and Mather, C. (1990) 'Photo essay: The Rio Grande borderland', *Journal of Cultural Geography*, 10(2): 66–98.

Witt, S.F. (1991) 'Tourism in Cyprus: Balancing the benefits and costs', *Tourism Management*, 12: 37–46.

Witt, S.F. (1998) 'Opening of the former communist countries of Europe to inbound tourism', in W.F. Theobald (ed.) *Global Tourism*, Second Edition, Oxford: Butterworth-Heinemann, 380–90.

Wolf, H.D. (1979) 'Der Einfluss der deutsch-österreichischen Staatsgrenze auf den Tourismus im Raum Salzburg, Bad Reichenhall, Berchtesgaden', in G. Gruber, H. Lamping, W. Lutz, J. Matznetter, and K. Vorlaufer (eds) *Tourism and Borders: Proceedings of the Meeting of the IGU Working Group - Geography of Tourism and Recreation*, Frankfurt: Institut für Wirtschafts- und Sozialgeographie der Johann Wolfgang Goethe Universität, 169–80.

Woodruffe, B.J. (1998) 'Conservation and the rural landscape', in D. Pinder (ed.) *The New Europe: Economy, Society and Environment*, Chichester: Wiley, 455–76.

Woolley, W. (1996) 'Border dispute hampers policing of center's lot', *Detroit News*, 30 October.

World Press Review (1993) 'Border crime', *World Press Review*, 40(6): 32.

World Tourism Organization (1988) *The Problems of Protectionism and Measures to Reduce Obstacles to International Trade in Tourism Services*, Madrid: WTO.

World Tourism Organization (1998 – updated) *Tourism to Benefit from Single European Currency*. Accessed 17 March 1998. http://www.world-tourism.org/pressrel/euro-a.htm

World Tourism Organization (2000a) *Compendium of Tourism Statistics, 1994–1998*. Madrid: WTO.

World Tourism Organization (2000b-updated) *Update*. Accessed 26 August 2000. http://www.world-tourism.org/

World Travel and Tourism Council (1991) *Bureaucratic Barriers to Travel*, Brussels: WTTC.

Wu, C. (1998) 'Cross-border development in Europe and Asia', *GeoJournal*, 44(3): 189–201.

Yenckel, J.T. (1995) 'Big bargains across the borders: Deals abound in Mexico and Canada as currencies plunge', *Washington Post*, 15 January: E2.

Yiftachel, O. (1990) 'Boundary change and institutional conflict in the planning of central Perth', *New Zealand Geographer*, 45: 8–67.

Young, L. and Rabb, M. (1992) 'New park on the bloc: Hungary, Czechoslovakia, and Austria must overcome obstacles to create Eastern Europe's first trilateral park', *National Parks*, 66(1): 35–40.

Yu, L. (1992) 'Emerging markets for China's tourism industry', *Journal of Travel Research*, 31(1): 10–13.

Yu, L. (1997) 'Travel between politically divided China and Taiwan', *Asia Pacific Journal of Tourism Research*, 2(1): 19–30.

Zelinski, W. (1988) 'Where every town is above average: Welcoming signs along America's highways', *Landscape*, 30(1): 1–10.

Zhang, G. (1993) 'Tourism crosses the Taiwan Straits', *Tourism Management*, 14: 228–31.

Zhao, X. (1994a) 'Barter tourism: A new phenomenon along the China–Russia border', *Journal of Travel Research*, 32(3): 65–7.

Zhao, X. (1994b) 'Barter tourism along the China–Russia border', *Annals of Tourism Research*, 21: 401–3.

Zhenge, P. (1993) 'Trade at the Sino–Russian border', *China Tourism*, 159: 70–3.

Index

Page references in *italics* indicate figures and tables, and those in **bold** indicate plates.